The Generalized Triangle Inequalities in Symmetric Spaces and Buildings with Applications to Algebra

MEMOIRS
of the
American Mathematical Society

Number 896

The Generalized Triangle Inequalities in Symmetric Spaces and Buildings with Applications to Algebra

Michael Kapovich
Bernhard Leeb
John J. Millson

March 2008 • Volume 192 • Number 896 (first of 5 numbers) • ISSN 0065-9266

American Mathematical Society
Providence, Rhode Island

2000 *Mathematics Subject Classification.*
Primary 22E46, 20G15; Secondary 14L24, 20E42.

Library of Congress Cataloging-in-Publication Data

Kapovich, Michael, 1963–
The generalized triangle inequalities in symmetric spaces and buildings with applications to algebra / Michael Kapovich, Bernhard Leeb, John J. Millson.
p. cm. — (Memoirs of the American Mathematical Society, ISSN 0065-9266 ; no. 896)
Includes bibliographical references.
ISBN 978-0-8218-4054-2 (alk. paper)
1. Semisimple Lie groups. 2. Linear algebraic groups. 3. Geometric group theory. 4. Lorentz groups. 5. Symmetric spaces. 6. Rings (Algebra) I. Leeb, Bernhard. II. Millson, John J. (John James), 1946– III. Title.

QA387.K366 2008
510 s—dc22
[512′.482] 2007060581

Memoirs of the American Mathematical Society

This journal is devoted entirely to research in pure and applied mathematics.

Subscription information. The 2008 subscription begins with volume 191 and consists of six mailings, each containing one or more numbers. Subscription prices for 2008 are US$675 list, US$540 institutional member. A late charge of 10% of the subscription price will be imposed on orders received from nonmembers after January 1 of the subscription year. Subscribers outside the United States and India must pay a postage surcharge of US$38; subscribers in India must pay a postage surcharge of US$43. Expedited delivery to destinations in North America US$53; elsewhere US$130. Each number may be ordered separately; *please specify number* when ordering an individual number. For prices and titles of recently released numbers, see the New Publications sections of the *Notices of the American Mathematical Society.*

Back number information. For back issues see the *AMS Catalog of Publications.*

Subscriptions and orders should be addressed to the American Mathematical Society, P. O. Box 845904, Boston, MA 02284-5904, USA. *All orders must be accompanied by payment.* Other correspondence should be addressed to 201 Charles Street, Providence, RI 02904-2294, USA.

Memoirs of the American Mathematical Society is published bimonthly (each volume consisting usually of more than one number) by the American Mathematical Society at 201 Charles Street, Providence, RI 02904-2294, USA. Periodicals postage paid at Providence, RI. Postmaster: Send address changes to Memoirs, American Mathematical Society, 201 Charles Street, Providence, RI 02904-2294, USA.

∞ The paper used in this book is acid-free and falls within the guidelines established to ensure permanence and durability.
Visit the AMS home page at http://www.ams.org/

10 9 8 7 6 5 4 3 2 1 13 12 11 10 09 08

Contents

Abstract

In this paper we apply our results on the geometry of polygons in infinitesimal symmetric spaces, symmetric spaces and buildings, [**KLM1, KLM2**], to four problems in algebraic group theory. Two of these problems are generalizations of the problems of finding the constraints on the eigenvalues (resp. singular values) of a sum (resp. product) when the eigenvalues (singular values) of each summand (factor) are fixed. The other two problems are related to the nonvanishing of the structure constants of the (spherical) Hecke and representation rings associated with a split reductive algebraic group over $\mathbb{Q}$ and its complex Langlands' dual. We give a new proof of the "Saturation Conjecture" for $GL(\ell)$ as a consequence of our solution of the corresponding "saturation problem" for the Hecke structure constants for all split reductive algebraic groups over $\mathbb{Q}$.

Received by the editor June 8, 2004.
2000 *Mathematics Subject Classification.* Primary 22E46, 20G15; Secondary 14L24, 20E42.
Key words and phrases. Symmetric spaces, Buildings, Hecke rings, Character rings.

CHAPTER 1

Introduction

In this paper we will examine and generalize the algebra problems listed immediately below from the point of view of spaces of non-positive curvature.

Let $\mathbb{F}$ be either the field $\mathbb{R}$ or $\mathbb{C}$, and let $\mathbb{K}$ be a nonarchimedean valued field with discrete valuation ring $\mathcal{O}$ and the value group $\mathbb{Z}$. For simplicity, let us consider here and throughout this Introduction a split reductive group $\underline{G}$ over $\mathbb{Q}$, see chapter 3 for a more general discussion.

- **Q1. Eigenvalues of a sum.** Set $G := \underline{G}(\mathbb{F})$, let K be a maximal compact subgroup of G. Let $\mathfrak{g}$ be the Lie algebra of G, and let $\mathfrak{g} = \mathfrak{k} + \mathfrak{p}$ be its Cartan decomposition. Give necessary and sufficient conditions on $\alpha, \beta, \gamma \in \mathfrak{p}/Ad(K)$ in order that there exist elements $A, B, C \in \mathfrak{p}$ whose projections to $\mathfrak{p}/Ad(K)$ are α, β and γ, respectively, so that
$$A + B + C = 0.$$
- **Q2. Singular values of a product.** Let G and K be the same as above. Give necessary and sufficient conditions on $\alpha, \beta, \gamma \in K\backslash G/K$ in order that there exist elements $A, B, C \in G$ whose projections to $K\backslash G/K$ are α, β and γ, respectively, so that
$$ABC = 1.$$
- **Q3. Invariant factors of a product.** Set $G := \underline{G}(\mathbb{K})$ and $K := \underline{G}(\mathcal{O})$. Give necessary and sufficient conditions on $\alpha, \beta, \gamma \in K\backslash G/K$ in order that there exist elements $A, B, C \in G$ whose projections to $K\backslash G/K$ are α, β and γ, respectively, so that
$$ABC = 1.$$
- **Q4. Decomposing tensor products.** Let $\underline{G}^\vee$ be the Langlands' dual group of $\underline{G}$, see Definition 2.12. Give necessary and sufficient conditions on highest weights α, β, γ of irreducible representations V_α, V_β, V_γ of $G^\vee := \underline{G}^\vee(\mathbb{C})$ so that
$$(V_\alpha \otimes V_\beta \otimes V_\gamma)^{G^\vee} \neq 0.$$

NOTATION 1.1. Throughout the paper we will denote by $Sol(\mathbf{Qi}, G)$ the sets of triples (α, β, γ) which are solutions of Problems **Qi**, $i = 1, ..., 4$ for the group G.

We refer the reader to section 3.5 for *linear algebra* reformulations of the problems **Q1, Q2, Q3** in the case when $\underline{G} = GL(m)$ and to section 8.2 for reformulation of **Q3** in terms of the Hecke ring.

The above problems have a long history which is described in detail in [**Fu1**], we briefly discuss it in section 3.5.

To state our main results we need several notation. The quotient spaces $\mathfrak{p}/Ad(K)$ and $K\backslash G/K$ in Problems **Q1** and **Q2** are naturally identified with the Euclidean Weyl chamber Δ inside a Cartan subspace $\mathfrak{a} \subset \mathfrak{p}$.

The double coset space appearing in Problem **Q3** is a more subtle discrete object, namely the intersection of Δ with the *cocharacter lattice* $L_{\underline{G}} \subset \mathfrak{a}$ (see section 3.4 for the precise definition). The appearance of the dual group $G^\vee$ in Problem **Q4** is explained by the fact that the cocharacter lattice of $\underline{G}$ is the same as the character lattice of the dual group $\underline{G}^\vee$, see section 3.4. We let Λ denote the lattice

$$\Lambda := \{(\alpha, \beta, \gamma) \in L_{\underline{G}}^3 : \alpha + \beta + \gamma \in Q(R^\vee)\}.$$

Here $Q(R^\vee)$ is the coroot lattice of $\underline{G}$. The importance of the lattice Λ comes from the inclusion

$$Sol(\mathbf{Q3}, \underline{G}(\mathbb{K})) \subset \Lambda.$$

Next we need the number k_R, the *saturation factor* associated with the root system R of the group $\underline{G}$; it is computed in the table 7.3 in section 7.1. For a split rank ℓ simple algebraic group $\underline{G}$ over $\mathbb{K}$ these numbers are defined as follows. Let R be the associated root system, let $\alpha_1, ..., \alpha_\ell$ be the simple roots and θ be the highest root:

$$\theta = \sum_{i=1}^{\ell} m_i \alpha_i. \tag{1.1}$$

Then k_R is the least common multiple of $m_1, ..., m_\ell$.

With these notation, the main result of this paper is the following chain of relations between the problems **Q1**–**Q4**:

MAIN THEOREM 1.2. *1.*

$$Sol(\mathbf{Q1}, \underline{G}(\mathbb{F})) = Sol(\mathbf{Q2}, \underline{G}(\mathbb{F})) \supset Sol(\mathbf{Q3}, \underline{G}(\mathbb{K})) \supset Sol(\mathbf{Q4}, \underline{G}^\vee(\mathbb{C})).$$

2.

$$Sol(\mathbf{Q2}, \underline{G}(\mathbb{F})) \cap k_R \cdot \Lambda \subset Sol(\mathbf{Q3}, \underline{G}(\mathbb{K})) \subset Sol(\mathbf{Q2}, \underline{G}(\mathbb{F})) \cap \Lambda$$

3. For the root systems $B_2 = C_2$ and G_2 we have

$$Sol(\mathbf{Q2}, \underline{G}(\mathbb{F})) \cap \Lambda \neq Sol(\mathbf{Q3}, \underline{G}(\mathbb{K}))$$

and

$$Sol(\mathbf{Q3}, \underline{G}(\mathbb{K})) \neq Sol(\mathbf{Q4}, \underline{G}^\vee(\mathbb{C})).$$

4. Even if $\underline{G}$ is not necessarily split, for $G = \underline{G}(\mathbb{F})$, the solution set $Sol(\mathbf{Q1}, G)$ depends only on the (finite) Weyl group of G. It is a convex homogeneous polyhedral cone $D_3(G/K)$ contained in Δ^3. A description of the linear inequalities defining this cone can be found in section 5.2.

5. Even if $\underline{G}$ is not necessarily split, for $G = \underline{G}(\mathbb{K})$, the solution set $Sol(\mathbf{Q3}, G)$ depends only on the affine Weyl group of G.

REMARK 1.3. The equivalence of Problems **Q1** and **Q2** (known as Thompson's Conjecture) for some classical groups (including $GL(m)$) was proved by Klyachko in [**Kly2**], for all complex semisimple groups by Alexeev, Meinrenken and Woodward in [**AMW**], and in general case by Evens and Lu in the recent paper [**EL**], using different methods.

REMARK 1.4. The above saturation constants k_R are not necessarily the smallest possible, for instance, in §8.7 we prove that for the group $\underline{G}$ with the root system G_2,

$$Sol(\mathbf{Q2}, \underline{G}(\mathbb{F})) \cap 2 \cdot \Lambda \subset Sol(\mathbf{Q3}, \underline{G}(\mathbb{K}))$$

REMARK 1.5. It is well-known, that

$$Sol(\mathbf{Q4}, \underline{G}(\mathbb{C})) \subset Sol(\mathbf{Q1}, \underline{G}(\mathbb{C})) \cap \Lambda,$$

see Theorem A.2 in the Appendix. In the forthcoming paper [**KM2**] it is shown that

$$Sol(\mathbf{Q3}, \underline{G}(\mathbb{K})) \cap k_R \cdot \Lambda \subset Sol(\mathbf{Q4}, \underline{G}(\mathbb{C})).$$

Despite of the algebraic appearance of the results of the Main Theorem, its main source is *geometry*, the sole exception is the relation between **Q3** and **Q4** which is proven via *Satake correspondence*. In this paper, we reformulate the algebra problems **Q1**–**Q3** as *geometric problems* which are special cases of a geometric question raised and studied in [**KLM1, KLM2**]:

We fix a Euclidean Coxeter complex (E, W_{aff}) with the Euclidean Weyl chamber $\Delta \cong E/W_{sph}$, and consider nonpositively curved metric spaces X with geometric structures *modeled on* (E, W_{aff}). The spaces of this kind which we are interested in are symmetric spaces of nonpositive curvature, their infinitesimal versions (*infinitesimal symmetric spaces*, see section 4.2), and Euclidean buildings. For such spaces X there is a notion of Δ-length for oriented geodesic segments which reflects the anisotropy of X. This leads to the following problem:

- **GTI: Generalized Triangle Inequalities.** Give necessary and sufficient conditions on $\alpha, \beta, \gamma \in \Delta$ in order that there exists a geodesic triangle in X with Δ-side lengths α, β and γ.

We refer to the solution set of **GTI** as $D_3(X)$.

As explained in section 4.4, for a symmetric space X, the problem **GTI** is equivalent to the Singular Value Problem **Q2**. For an infinitesimal symmetric space $\mathfrak{p}$, it is equivalent to the Eigenvalues of a Sum Problem **Q1**:

THEOREM 1.1. *Suppose that $\underline{G}$ is a reductive algebraic group over $\mathbb{R}$ and $\mathbb{F} = \mathbb{C}$ or $\mathbb{F} = \mathbb{R}$; let $X = G/K$ be the symmetric space corresponding to $G = \underline{G}(\mathbb{F})$ and $\mathfrak{p}$ be the infinitesimal symmetric space. Then*

$$Sol(\mathbf{Q1}, G) = D_3(\mathfrak{p}), \quad Sol(\mathbf{Q2}, G) = D_3(X).$$

The Invariant Factor Problem **Q3** corresponds to the case when X is a Euclidean building and in this case the inequalities defining $D_3(X)$ give only *necessary* conditions on (α, β, γ) to solve the algebraic problem **Q3**. The key difference with the problems **Q1, Q2** is lack of homogeneity in Euclidean buildings X: *Not all vertices of X are special* (except for the buildings of type A_ℓ). This eventually explains why for the Euclidean building X corresponding to some groups $\underline{G}(\mathbb{K})$ we have the inequality

$$D_3(X) \cap \Lambda \neq Sol(\mathbf{Q3}, \underline{G}(\mathbb{K})).$$

We refer the reader to section 4.4 for justification of the correspondence between algebraic and geometric problems.

Despite of the results of this paper, the structure of the solution sets $Sol(\mathbf{Q3})$, $Sol(\mathbf{Q4})$ remains unclear. It was known for a long time that $Sol(\mathbf{Q4})$ is a semigroup.

In contrast, we will see in section 8.7 that the set $Sol(\mathbf{Q3})$ is not a semigroup for the root systems $B_2 = C_2$ and G_2.

The following result, a consequence of a recent theorem in logic of M. C. Laskowski [**La**], building on work of S. Kochen [**Ko**], reveals the general structure of the subset

$$Sol(\mathbf{Q3}, \underline{G}(\mathbb{K})) \subset L^3_{\underline{G}}.$$

Define a subset of $L^3_{\underline{G}}$ to be *elementary* if it is the set of solutions of a finite set of linear inequalities with integer coefficients and a finite set of linear congruences. Then Laskowski proves the following

THEOREM 1.2 (M. C. Laskowski, [**La**]). *There is an integer $N = N_{\underline{G}}$, depending only on $\underline{G}$, such that for any nonarchimedean Henselian valued field $\mathbb{K}$ with value group $\mathbb{Z}$ and residue characteristic greater than N we have*

$$Sol(\mathbf{Q3}, \underline{G}(\mathbb{K})) \subset L^3_{\underline{G}} \text{ is a finite union of elementary sets.}$$

REMARK 1.6. It follows from Theorem 1.2, Part 5, that Theorem 1.2 holds for *all* complete nonarchimedean valued fields $\mathbb{K}$. This is because all split groups $\underline{G}$, as $\mathbb{K}$ varies, have the same affine Coxeter group. Hence the set $Sol(\mathbf{Q3}, \underline{G}(\mathbb{K}))$ does not depend on $\mathbb{K}$. Therefore, once the above statement is true for one of them it is true for all of them.

PROBLEM 1.7. 1. Find the corresponding inequalities and congruences.
2. Is there an analogue of Laskowski's theorem that describes the set Sol($\mathbf{Q4}$)? If so find the inequalities and congruences in this case as well.

We can solve the Problem 1.7 and thus the Problem **Q3** for the following groups:

THEOREM 1.3. *If $\underline{G}$ is covered by $SL(m)$ or $GL(m)$ then*

$$Sol(\mathbf{Q3}, \underline{G}(\mathbb{K})) = Sol(\mathbf{Q1}, \underline{G}(\mathbb{F})) \cap \Lambda.$$

In other words, say for the group $\underline{G} = SL(m)$, the solution set $Sol(\mathbf{Q3}, SL(m, \mathbb{K}))$ is the set of triples (α, β, γ) in $D_3(Sym_m) \cap \mathbb{Z}^{3m}$ such that $\alpha + \beta + \gamma \in Q(R^\vee)$. Here Sym_ℓ is the infinitesimal symmetric space of zero trace symmetric $m \times m$ matrices and $Q(R^\vee)$ is the coroot lattice of $SL(m)$.

A similar statement holds for a more general class of groups, see Theorem 7.4. For example, the conclusion of part (a) holds for the groups $SL(\ell, \mathbb{D})$ where $\mathbb{D}$ is a division algebra over $\mathbb{K}$, see Tableau des échelonages [**BT**, p. 29-30].

We conclude by pointing out that for the group $GL(m)$ it was known since the 1960-s that

$$Sol(\mathbf{Q3}, GL(m)) = Sol(\mathbf{Q4}, GL(m))$$

and therefore we can use Theorem 1.3 to give a new proof of the Saturation Conjecture for $GL(m)$ (the theorem of Knutson and Tao), see section 8.6.

This paper is organized as follows.

In chapter 2 we review root systems for algebraic reductive groups and discuss Coxeter groups.

In chapter 3 we set up the general algebra problem **R** which generalizes the setting of the Problems **Q1**–**Q3**. We then discuss in detail the parameter spaces for the Problems **Q1**–**Q4** and their mutual relation.

In chapter 4 we first convert the problem **R** into an abstract geometry problem about existence of polygons with the prescribed generalized side-lengths (Problem

4.1). Next, we introduce a class of metric spaces (metric spaces modeled on Euclidean Coxeter complexes) and restate the abstract geometry problem for three classes of such spaces: infinitesimal symmetric spaces, nonpositively curved symmetric spaces and Euclidean buildings. We introduce the notion of refined and coarse (the Δ-length) generalized side-lengths for the geodesic polygons in such metric spaces. For the infinitesimal symmetric spaces and symmetric spaces, the problem of existence of polygons with the prescribed Δ-side lengths is adequate for solving the corresponding algebra problems (**Q1** and **Q2**), but in the case of buildings it is not. This establishes the equalities and inclusions

$$Sol(\mathbf{Q3}) \subset Sol(\mathbf{Q1}) = Sol(\mathbf{Q2}) = D_3(X)$$

in Theorem 1.2 as well as Part 4 and 5 in this theorem.

In chapter 5 we describe the solution (given in [**KLM1, KLM2**]) to the problem **GTI** of existence of polygons with the prescribed Δ-side lengths in the above classes of metric spaces X. We also discuss the relation of this solution to symplectic and Mumford quotients. We describe the system of generalized triangle inequalities for X and give an explicit set of inequalities in the case of root system of type B_2. *The reader who is not interested in the geometric ideas behind the results of this paper can skip this chapter except for the definition of the mapping Φ_ψ for buildings which will be used in chapter 7.*

In chapter 6 we show that the geometry problem **GTI** solved in the previous chapter is not adequate (in the building case) for solving the algebra problem **Q3**. The key tool here is the procedure of *folding* of geodesic triangles in Euclidean buildings into apartments. The main application of the examples constructed in this section is to establish the first inequality in Part 3 of Theorem 1.2.

In chapter 7 we show that in the case of A-type root systems, the solution of the geometry problem **GTI** given in chapter 5 solves the algebra problem **Q3** as well:

$$Sol(\mathbf{Q3}) = D_3(X) \cap \Lambda.$$

More generally, we establish the existence of the saturation factors $k = k_{inv\ fact}$ and compute these numbers; modulo multiplication by k the geometry problem **GTI** is equivalent to the algebra problem **Q3**. This establishes Part 2 of Theorem 1.2.

In chapter 8 we reformulate problem **Q3** in terms of *spherical Hecke rings*, compare the algebraic problems **Q3** and **Q4**, establish the inclusion

$$Sol(\mathbf{Q4}, \underline{G}^\vee(\mathbb{C})) \subset Sol(\mathbf{Q3}, \underline{G}(\mathbb{K}))$$

(see Part 1 of Theorem 1.2) and give our new proof of the saturation conjecture for $GL(m)$.

In the Appendix A we relate the solution set of the problem **Q4** to Mumford quotients and show that the former forms a semigroup. Although these results are known to experts, we include them for the sake of completeness.

Acknowledgments. During the work on this paper the first author was visiting the Max Plank Institute (Bonn), he was also supported by the NSF grants DMS-99-71404 and DMS-02-03045. The first and the third authors were supported by the Mathematics Department of the University of Tübingen during their stay there in May of 2002. The third author was supported by the NSF grant DMS-01-04006. The authors gratefully acknowledge support of these institutions. We are grateful the referees of this paper for their useful remarks. The authors are grateful to L. Ein, S. Kumar, T. Haines, R. Hemmecke, C. Laskowski, J. Stembridge, E. Vinberg and C. Woodward for useful conversations. We would especially like to thank Steve Kudla for suggesting we use the Satake transform to connect Problems **Q3** and **Q4** and Jiu-Kang Yu for his help with chapter 8 and in particular for providing us with the proof of Lemma 8.15.

When we told G. Lusztig of our Theorem 8.5 establishing the inclusion

$$Sol(\mathbf{Q4}, \underline{G}^\vee(\mathbb{C})) \subset Sol(\mathbf{Q3}, \underline{G}(\mathbb{K})),$$

for local nonarchimedean valued fields $\mathbb{K}$, he informed us that although he had not known the result before our message, it was an easy consequence of his work in [**Lu2**]. We should say that our proof depends in an essential way on Lusztig's paper [**Lu2**]. We use his change of basis formulas, Lemma 8.11, and his realization that the coefficients in one of those formulas were Kazhdan-Lusztig polynomials for the affine Weyl group, Lemma 8.12. Subsequently, alternative proofs of Theorem 8.5 were given by Tom Haines in [**Ha2**] and by Kapovich and Millson in [**KM2**].

CHAPTER 2

Roots and Coxeter Groups

Let $\underline{G}$ be a reductive algebraic group over a field $\mathbb{F}$ and $\underline{T}$ be a split torus in $\underline{G}$. Our goal in this section is to describe *the root datum* associated to the pair $(\underline{G}, \underline{T})$. The reader will find the definition of root datum in [**Sp**, §1].

2.1. Split tori over $\mathbb{F}$

We recall that the algebraic group $\mathbb{G}_m$ is the affine algebraic group with coordinate ring $\mathbb{F}[S,T](ST-1)$ and comultiplication Δ given by

$$\Delta(T) = T \otimes T, \quad \Delta(S) = S \otimes S.$$

DEFINITION 2.1. An affine algebraic group $\underline{T}$ defined over $\mathbb{F}$ is a split torus of rank l if it is isomorphic to the product of l copies of $\mathbb{G}_m$.

A character of an algebraic group $\underline{T}$ defined over $\mathbb{F}$ is a morphism of algebraic groups from $\underline{T}$ to $\mathbb{G}_m$. The product of two characters and the inverse of a character are characters and accordingly the set of characters of $\underline{T}$ is an abelian group denoted by $X^*(\underline{T})$.

LEMMA 2.2. *Suppose that $\underline{T}$ is a split torus over $\mathbb{F}$ of rank l. Then the character group of $\underline{T}$ is a lattice (i.e. free abelian group) of rank l.*

Proof: We have $\mathbb{F}[\underline{T}] \cong \mathbb{F}[T_1, T_1^{-1}, \cdots, T_l, T_l^{-1}]$. A character of $\underline{T}$ corresponds to a Hopf algebra morphism from $\mathbb{F}[\mathbb{G}_m]$ to $\mathbb{F}[T_1, T_1^{-1}, \cdots, T_l, T_l^{-1}]$. Such a morphism is determined by its value on T. This value is necessarily a grouplike element (this means $\Delta(f) = f \otimes f$). However the grouplike elements of $\mathbb{F}[T_1, T_1^{-1}, \cdots, T_l, T_l^{-1}]$ are the monomials in the T_i's and their inverses. The exponents of the monomial give the point in the lattice. □

COROLLARY 2.3.

$$Hom(\mathbb{G}_m, \mathbb{G}_m) \cong \mathbb{Z}.$$

We note that the previous isomorphism is realized as follows. Any Hopf-algebra homomorphism ψ of coordinate rings is of the form $T \to T^n$ for some integer n. Then the above isomorphism sends ψ to n.

DEFINITION 2.4. A cocharacter or a one-parameter (algebraic) subgroup of $\underline{T}$ is a morphism $\phi : \mathbb{G}_m \to T$. The set of cocharacters of $\underline{T}$ will be denoted $X_*(\underline{T})$.

LEMMA 2.5. *Suppose that $\underline{T}$ is a split torus over $\mathbb{F}$ of rank l. Then the cocharacter group of $\underline{T}$ is a free abelian group of rank l.*

Proof: We have $\mathbb{F}[\underline{T}] \cong \mathbb{F}[T_1, T_1^{-1}, \cdots, T_l, T_l^{-1}]$. A cocharacter of $\underline{T}$ corresponds to a Hopf algebra morphism ψ from $\mathbb{F}[T_1, T_1^{-1}, \cdots, T_l, T_l^{-1}]$ to $\mathbb{F}[\mathbb{G}_m]$. Such a morphism is determined by its value on $T_1, \cdots, T_l$. Then ψ corresponds to the lattice vector $(m_1, \cdots, m_l)$ where $T^{m_i} = \psi(T_i)$. □

We define an integer-valued pairing $\langle\ ,\ \rangle$ between characters and cocharacters as follows. Suppose ϕ is a cocharacter of $\underline{T}$ and χ is a character. Then $\chi \circ \phi \in Hom(\mathbb{G}_m, \mathbb{G}_m) \cong \mathbb{Z}$. We define $\langle \chi, \phi \rangle$ to be the integer corresponding to $\chi \circ \phi$.

We now describe two homomorphisms that will be useful in what follows. Let $T_e(\underline{T})$ be the Zariski tangent space of $\underline{T}$ at the identity e.

DEFINITION 2.6. We define $\Phi : X_*(\underline{T}) \longrightarrow T_e(\underline{T})$ by

$$\Phi(\lambda) = \lambda'(1).$$

Here 1 is the identity of $GL(1, \mathbb{F})$ and $\lambda'(1)$ denotes the derivative at 1.

We also define $\Phi^\vee : X^*(\underline{T}) \longrightarrow T_e^*(\underline{T})$ by

$$\Phi^\vee(\lambda) = d\lambda|_e.$$

REMARK 2.7. The character and cocharacter groups $X^*(\underline{T}), X_*(\underline{T})$ are multiplicative groups, the trivial (co)character will be denoted by 1. However, we will use the embeddings Φ and $\Phi^\vee$ to identify them with additive groups. This will be done for the most part in chapters 3 and 8.

2.2. Roots, coroots and the Langlands' dual

The reductive group $\underline{G}$ picks out a distinguished finite subset of $X^*(\underline{T})$, the *relative root system* $R = R_{rel}(\underline{G}, \underline{T})$. A character of $\underline{T}$ is a *root* if it occurs in the restriction of the adjoint representation of $\underline{G}$ to $\underline{T}$. We let $Q(R)$ denote the subgroup of $X^*(\underline{T})$ generated by R and define $V := Q(R) \otimes \mathbb{R}$.

We recall

DEFINITION 2.8. The algebraic group $\underline{G}$ is *split over* $\mathbb{F}$ if it has a maximal torus $\underline{T}$ defined over $\mathbb{F}$, which is split.

From now on we assume $\underline{G}$ is split over $\mathbb{F}$ and $\underline{T}$ is a maximal torus as in the above definition.

It is proved in [**Sp**, §2], that $R \subset V$ satisfies the axioms of a root system. Moreover in the same section it is proved that to every root $\alpha \in R$ there is an associated coroot $\alpha^\vee \in X_*(\underline{T})$ such that $\langle \alpha, \alpha^\vee \rangle = 2$. We let $R^\vee$ denote the resulting set of coroots, let $Q(R^\vee)$ be the subgroup of $X_*(\underline{T})$ they generate and $V^\vee := Q(R^\vee) \otimes \mathbb{R}$. The root and coroot system R and $R^\vee$ determine (isomorphic) finite Weyl groups $W, W^\vee$ which acts on $V^\vee$ and V respectively. The action of the generators $s_\alpha, s_{\alpha^\vee}$ of the group $W, W^\vee$ on $X^*(\underline{T})$ and $X_*(\underline{T})$ are determined by the formulae:

$$s_\alpha(x) := x - \langle x, \alpha^\vee \rangle \alpha \text{ and } s_{\alpha^\vee}(u) := u - \langle u, \alpha \rangle \alpha^\vee.$$

We then have [**Sp**, §2]:

PROPOSITION 2.9. *The quadruple* $\Psi(\underline{G}, \underline{T}) := (X^*(\underline{T}), R, X_*(\underline{T}), R^\vee)$ *is a* root datum.

DEFINITION 2.10. Let $\Psi = (X, R, X^\vee, R^\vee)$ and $\Psi' = (X', R', (X')^\vee, (R')^\vee)$ be root data. Then an *isogeny* from Ψ' to Ψ is a homomorphism ϕ from X' to X such that ϕ is injective with finite cokernel. Moreover we require that ϕ induces a bijection from R' to R and the transpose of ϕ induces a bijection of coroots.

Now suppose $f : \underline{G} \longrightarrow \underline{G}'$ is a covering of algebraic groups. If $\underline{T}$ is a maximal torus in $\underline{G}$ then its image $\underline{T}'$ is a maximal torus in $\underline{G}'$. The induced map on characters gives rise to an isogeny of root data, denoted $\Psi(f)$.

Conversely, we have [**Sp**, Theorem 2.9]:

THEOREM 2.1. (1) *For any root datum Ψ with reduced root system there exist a connected split (over $\mathbb{F}$) reductive group $\underline{G}$ and a maximal split torus $\underline{T}$ such that $\Psi = \Psi(\underline{G}, \underline{T})$. The pair $(\underline{G}, \underline{T})$ is unique up to isomorphism.*
(2) *Let $\Psi = \Psi(\underline{G}, \underline{T})$ and $\Psi' = \Psi(\underline{G}', \underline{T}')$ and ϕ be an isogeny from Ψ' to Ψ. Then there is a covering $f : \underline{G} \longrightarrow \underline{G}'$ with the image of $\underline{T}$ equal to $\underline{T}'$ such that $\phi = \Psi(f)$.*

Before stating the next definition we need a lemma which we leave to the reader.

LEMMA 2.11. *If $(X, R, X^\vee, R^\vee)$ is a root datum so is $(X^\vee, R^\vee, X, R)$.*

We now have

DEFINITION 2.12. Let $\underline{G}$ be a (connected) split reductive group over $\mathbb{Q}$. Let $\Psi = \Psi(\underline{G}, \underline{T}) = (X, R, X^\vee, R^\vee)$ be the root datum of $(\underline{G}, \underline{T})$. Then the *Langlands dual* $\underline{G}^\vee$ of $\underline{G}$ is the unique (up to isomorphism) reductive group over $\mathbb{Q}$ which has the root datum $\Psi' = (X^\vee, R^\vee, X, R)$.

In fact we will need only the complex points $G^\vee := \underline{G}^\vee(\mathbb{C})$ of $\underline{G}^\vee$ in what follows. We will accordingly abuse notation and frequently refer to $G^\vee$ as the Langlands dual of $\underline{G}$. One has

$$(\underline{G}^\vee)^\vee \cong \underline{G}.$$

2.3. Coxeter groups

In this section we review the properties of Coxeter groups, we refer the reader for a more thorough discussion to [**Hum**, Section 4.2] and [**Bo**]. Let $R \subset V^*$ be a root system of rank n on a real Euclidean vector space V. We do not assume that n equals the dimension of V. Note that in the case of semisimple Lie algebras, the space V will be a Cartan subalgebra $\mathfrak{a} \subset \mathfrak{g}$ with the Killing form.

We will sometimes identify V and V^* using the metric. Let $Q(R) \subset V^*$ denote *the root lattice*, i.e. the integer span of R. This subgroup is a lattice in $Span_{\mathbb{R}}(R) \subset V$, it is a discrete free abelian subgroup of rank n. Given a subgroup $\Lambda \subset \mathbb{R}$ we define a collection $\mathcal{H} = \mathcal{H}_{R,\Lambda}$ of hyperplanes (called *walls*) in V as the set

$$\mathcal{H} := \{H_{\alpha,t} = \{v \in V : \alpha(v) = t\}, t \in \Lambda, \alpha \in R\}.$$

In this paper we will be mostly interested in the case when Λ is either $\mathbb{Z}$ or $\mathbb{R}$, but much of our discussion is more general. We define an *affine Coxeter group* $W_{aff} = W_{R,\Lambda}$ as the group generated by the reflections w_H in the hyperplanes $H \in \mathcal{H}$. The only reflection hyperplanes of the reflections in W_{aff} are the elements of $\mathcal{H}$. The pair (E, W_{aff}) is called a *Euclidean Coxeter complex*, where $E = V$ is the Euclidean space. The *vertices* of the Coxeter complex are points which belong to the transversal intersections of n walls in $\mathcal{H}$. (This definition makes sense even if $n < \dim(V)$, only in this case there will be continuum of vertices even if $\Lambda = \mathbb{Z}$.) If W_{aff} is trivial, we declare *each point* of E a vertex. We let $E^{(0)}$ denote the vertex set of the Coxeter complex.

DEFINITION 2.13. An *embedding of Euclidean Coxeter complexes* is a map $(f, \phi) : (E, W) \to (E', W')$, where $\phi : W \to W'$ is a monomorphism of Coxeter groups and $f : E \to E'$ is a ϕ-equivariant affine embedding.

Let L_{trans} denote the translational part of W_{aff}. If $\Lambda = \mathbb{Z}$ then L_{trans} is the *coroot lattice* $Q(R^\vee)$ of R. In general, $L_{trans} = Q(R^\vee) \otimes \Lambda$. The linear part W_{sph} of W_{aff} is a finite Coxeter group acting on V, it is called a *spherical Coxeter group.* The stabilizer of the origin $0 \in E$ (which we will regard as a base-point $o \in E$) in W_{aff} maps isomorphically onto W_{sph}. Thus $W_{aff} = W_{sph} \ltimes L_{trans}$. A vertex of the Euclidean Coxeter complex is called *special* if its stabilizer in W_{aff} is isomorphic to W_{sph}. We let $E^{(0),sp}$ denote the set of special vertices of E.

REMARK 2.14. The normalizer N_{aff} of W_{aff} (in the full group V of translations on E) acts transitively on the set of special vertices. The vertex set $E^{(0)}$ of the complex (E, W_{aff}) contains $N_{aff} \cdot o$, but typically it is strictly larger that. Moreover, in many cases $E^{(0)}$ does not form a group.

We recall that the weight group $P(R)$ and the coweight group $P(R^\vee)$ are defined by

$$\begin{aligned} P(R) &= \{\lambda \in V^* : \lambda(v) \in \mathbb{Z}, \forall v \in R^\vee\}, \\ P(R^\vee) &= \{v \in V : \alpha(v) \in \mathbb{Z}, \forall \alpha \in R\}. \end{aligned}$$

REMARK 2.15. In the case when $n < \dim(V)$ our definition of weights is different from the one in [**Sp**].

Again, $P(R)$ and $P(R^\vee)$ are lattices provided that $n = \dim(V)$, otherwise they are nondiscrete abelian subgroups of V. We have the inclusions

$$Q(R^\vee) \subset P(R^\vee), \quad Q(R) \subset P(R).$$

The normalizer N_{aff} equals $P(R^\vee) \otimes \Lambda$.

The spherical Coxeter groups W_{sph} which appear in the above construction act naturally on the sphere at infinity $S = \partial_\infty E$; the pair (S, W_{sph}) is called a *spherical Coxeter complex*. The definitions of *walls*, *vertices*, etc., for Euclidean Coxeter complexes generalize verbatum to the spherical complexes. We will use the notation $\Delta_{sph} \subset S$ for the spherical Weyl chamber, Δ_{sph} is the ideal boundary of the Euclidean Weyl chamber $\Delta \subset E$ (i.e. a fundamental domain for the action $W_{sph} \curvearrowright E$, which is bounded by walls).

From our viewpoint, the Euclidean Coxeter complex is a more fundamental object than a root system. Thus, if the root system R was not reduced, we replace it with a reduced root system R' which has the same group W_{aff}: If $\alpha, 2\alpha \in R$ we retain the root 2α and eliminate the root α. We will assume henceforth that the root system R is reduced.

Product decomposition of Euclidean Coxeter complexes. Suppose that (E, W_{aff}) is a Euclidean Coxeter complex associated with the reduced root system R, let $R_1, ..., R_s$ denote the decomposition of R into irreducible components. Accordingly, the Euclidean space E splits as the metric product

$$E = E_0 \times \prod_{i=1}^{s} E_i,$$

where E_i is spanned by $R_i^\vee$, $1 \leq i \leq s$. This decomposition is invariant under the group W_{aff} which in turn splits as

$$W_{aff} = \prod_{i=1}^{s} W_{aff}^i,$$

where $W_{aff}^i = W_{R_i,\Lambda}$ for each $i = 1, ..., s$; for $i = 0$ we get the trivial Coxeter group W_{aff}^0. The group W_{aff}^i is the image of $W_{R_i,\Lambda}$ under the natural embedding of affine groups

$$Aff(E_i) \longrightarrow \prod_{i=1}^{s} Aff(E_i) = Aff(E).$$

Analogously, the spherical Coxeter group W_{sph} splits as the direct product

$$W_{sph}^0 \times ... \times W_{sph}^s$$

(where $W_{sph}^0 = \{1\}$). The Weyl chamber Δ of W_{sph} is the direct product of the Weyl chambers $\Delta_0 \times \Delta_1 \times ... \times \Delta_s$, where Δ_i is a Weyl chamber of W_{sph}^i and $\Delta_0 = E_0$. Similarly, the normalizer N_{aff} of W_{aff} splits as

$$N_{aff} = V_0 \times \prod_{i=1}^{s} N_{aff}^i,$$

where V_0 is the vector space underlying E_0 and N_{aff}^i is the normalizer of W_{aff}^i in the group of translations of E_i. Note that for each $i = 1, ..., s$ the groups W_{aff}^i and N_{aff}^i act as lattices on E_i. We observe that the vertex set of the complex (E, W_{aff}) equals

$$E_0 \times E_1^{(0)} \times ... \times E_s^{(0)},$$

where $E_i^{(0)}$ is the vertex set of the complex (E_i, W_{aff}^i). Similarly, the set of special vertices $E^{(0),sp}$ of E equals

$$E_0 \times E_1^{(0),sp} \times ... \times E_s^{(0),sp}.$$

REMARK 2.16. Our discussion of Coxeter groups was somewhat nongeometric; a more geometric approach would be to start with an affine Coxeter group and from this get root systems, etc.

CHAPTER 3

The First Three Algebra Problems and the Parameter Spaces Σ for $K\backslash\overline{G}/K$

We will see in this chapter that the problems **Q1**–**Q3** for reductive algebraic groups G can reformulated as special cases of a single algebraic problem as follows. There is a group $\overline{G}$ (closely associated to G) which contains K, a maximal bounded subgroup of G. The conditions of fixing α, β and γ in problems **Q1**–**Q3** will amount to fixing three double cosets in $\Sigma = K\backslash\overline{G}/K$. The problems **Q1**–**Q3** will be then reformulated as:

- **Problem R($\overline{G}$)**: Find necessary and sufficient conditions on $\alpha, \beta, \gamma \in \Sigma$ in order that there exist $A, B, C \in \overline{G}$ in the double cosets represented by α, β, γ resp., such that $A \cdot B \cdot C = 1$.

We will now describe the groups $\overline{G}$ and K for the problems **Q1**–**Q3**. The main part of this chapter will then be occupied with describing the double coset spaces $\Sigma = K\backslash\overline{G}/K$. In section 3.1 we will also prove that the problem **R($\overline{G}$)** agrees with the Problem **Q1** from the Introduction (the equivalence will be clear for two other problems).

(1) For the Problem **Q1**: For $\mathbb{F} = \mathbb{R}$ or $\mathbb{C}$, let $\underline{G}$ be a connected reductive algebraic group over $\mathbb{R}$, $G := \underline{G}(\mathbb{F})$ be a real or complex Lie group with Lie algebra $\mathfrak{g}$. Pick a maximal compact subgroup K of G. Let $\mathfrak{k}$ denote the Lie algebra of K. Then we have the orthogonal decomposition (with respect to the Killing form)

$$\mathfrak{g} = \mathfrak{k} \oplus \mathfrak{p}.$$

We let $\overline{G}$ be the *Cartan motion group* $\overline{G} = K \ltimes \mathfrak{p}$.

(2) For the problem **Q2**: $\underline{G}$ and K are the same as in 1, but now we take $\overline{G} = G = \underline{G}(\mathbb{F})$.

(3) For the problem **Q3**: Let $\mathbb{K}$ be a complete nonarchimedean valued field with a discrete valuation v and the value group $\mathbb{Z} = v(\mathbb{K}) \subset \mathbb{R}$. Let $\mathcal{O}$ denote the ring of elements in $\mathbb{K}$ with nonnegative valuation. Let $\underline{G}$ be a connected reductive algebraic group over $\mathbb{K}$, $\overline{G} := G := \underline{G}(\mathbb{K})$ and $K := G(\mathcal{O})$.

In order to relate Problems **Q1**–**Q3** with geometry we will have to compute the double coset spaces $\Sigma = K\backslash\overline{G}/K$. Moreover we will show that the parameter spaces Σ admit canonical linear structures: In the first two problems we use a Cartan decomposition $G = KAK$, then Σ is described as the (positive) Weyl chamber Δ in $\mathfrak{a}$, the Lie algebra of A. In the case of the third problem, Σ is identified with the intersection

$$L_{\underline{G}} \cap \Delta \subset \mathfrak{a} = X_*(\underline{T}) \otimes \mathbb{R},$$

where $\underline{T}$ is a maximal $\mathbb{K}$-split torus of $\underline{G}$ and $L_{\underline{G}}$ is *extended cocharacter lattice* of $\underline{T}$ (see section 3.3).

3.1. The generalized eigenvalues of a sum problem Q1 and the parameter space Σ of K-double cosets

We continue with the notation of previous section, in particular, $\overline{G} = K \ltimes \mathfrak{p}$. We now describe the parameter space $\Sigma = K\backslash\overline{G}/K$. We choose a Cartan subspace $\mathfrak{a} \subset \mathfrak{p}$ (i.e. a maximal subalgebra of $\mathfrak{p}$, necessarily abelian and reductive). We recall that the positive Euclidean Weyl chamber Δ is the dual cone in $\mathfrak{a}$ to the cone of positive restricted roots in $\mathfrak{a}^*$.

LEMMA 3.1. *The inclusion $\iota : \Delta \to K \ltimes \mathfrak{p}$ given $\iota(\sigma) = (1, \sigma)$ induces a bijection onto Σ.*

Proof: We first observe that since $(k_1, x) \cdot (k_2, 0) = (k_1k_2, x)$, every double coset in $K\backslash\overline{G}/K$ has a representative of the form $(1, x)$. Since

$$(k, 0) \cdot (1, x) \cdot (k, 0)^{-1} = (1, Ad\, k(x))$$

the lemma is clear. □

We complete the proof of equivalence of $\mathbf{R}(\overline{G})$ with Problem **Q1** by observing that for $g_i = (k_i, x_i)$, $i = 1, 2, 3$,

$$(k_1, x_1) \cdot (k_2, x_2) \cdot (k_3, x_3) = (k_1k_2k_3, x_1 + Adk_1x_2 + Ad(k_1k_2)x_3). \tag{3.1}$$

We see from (3.1) that if $g_1g_2g_3 = 1$ then putting

$$a = x_1, \quad b = Adk_1x_2, \quad c = Ad(k_1k_2)x_3$$

we find

$$a + b + c = 0.$$

Thus we find representatives a, b and c in $\mathfrak{p}$ of the orbits $Ad(K)(x_i), 1 \le i \le 3$ whose sum equals zero (i.e. we have a solution of the Eigenvalues of a Sum Problem). Conversely, if $a, b, c \in \mathfrak{p}$ solve the Eigenvalues of a Sum Problem then

$$g_1 \cdot g_2 \cdot g_3 = (1, a) \cdot (1, b) \cdot (1, c) = (1, 0)$$

and hence g_1, g_2, g_3 solve the double coset problem $\mathbf{R}(\overline{G})$.

Thus we may find A, B and C in $\overline{G}$ in the required K-double cosets with $A \cdot B \cdot C = 1$ if and only if there exist $a, b, c \in \mathfrak{p}$ with $a + b + c = 0$ in the required $Ad\, K$-orbits in $\mathfrak{p}$.

3.2. The generalized singular values of a product and the parameter space Σ of K-double cosets

Let $\overline{G} := G = \underline{G}(\mathbb{F})$ and consider the Cartan decomposition $G = KAK$ where A is the maximal abelian subgroup of $\overline{G}$, whose Lie algebra is the Cartan subalgebra $\mathfrak{a} \subset \mathfrak{p}$ from the previous section. We will identify A and $\mathfrak{a}$ via the exponential map, let $A_\Delta := \exp(\Delta) \subset A$. Then we get the refinement of the Cartan decomposition: $G = KA_\Delta K$. It is now clear that the inclusion $\iota : \Delta \to G$ given by $\iota(\sigma) = \exp(\sigma)$ induces a bijection onto $\Sigma = K\backslash G/K$. In what follows we will parameterize double cosets in Σ by their logarithms, which are vectors in Δ.

3.3. The generalized invariant factor problem and the parameter space Σ of K-double cosets

The goal of this section is to give an explicit description of the parameter space $\Sigma = \Sigma_{inv\ fact}$ for the Invariant Factor Problem **Q3**. We start by noticing that the discrete valuation $v : \mathbb{K} \to \mathbb{Z}$ admits a splitting $\kappa : \mathbb{Z} \to \mathbb{K}$: it is given by sending $1 \in \mathbb{Z}$ to the uniformizer $\pi \in \mathbb{K}$. Throughout this section, $\underline{G}$ will denote a connected reductive algebraic group over $\mathbb{K}$, which has the (relative) rank l. Let $\underline{T} \subset \underline{G}$ be a maximal $\mathbb{K}$-split torus, and let $X^*(\underline{T})$ and $X_*(\underline{T})$ be the lattices of characters and cocharacters. We denote by $\underline{P} \subset \underline{G}$ be a minimal parabolic subgroup such that $(\underline{P}, \underline{T})$ is a parabolic pair, [**C**, pg. 127].

This data determines a (relative) root system R_{rel} and hence a (relative) finite Weyl group W_{sph}. Let $\underline{Z} \subset \underline{G}$ and $\underline{N} \subset \underline{G}$ be respectively the centralizer and normalizer (over $\mathbb{K}$) of the algebraic group $\underline{T}$. As usual we let G, N, Z and T denote the groups of $\mathbb{K}$-points of the corresponding algebraic groups. We let K denote $\underline{G}(\mathcal{O})$ and $B := \underline{T}(\mathcal{O}) = K \cap T$. We will frequently refer to G as a *nonarchimedean reductive Lie group*.

As in the previous section, we have to discuss the Cartan decomposition $G = K \cdot \Gamma \cdot K$ of the group G (where Γ will be an appropriate subset of Z). In the split case this decomposition will have the form $G = K \cdot A_\Delta \cdot K$, where $A_\Delta \subset A$ and A will be a subgroup of T so that $A \times B = T$.

We can now describe $\Sigma_{inv\ fact}(G) = K\backslash G/K$, the parameter space for the generalized invariant factors problem for the nonarchimedean reductive Lie group G.

Split Case. We first consider the case when $\underline{G}$ is split over $\mathbb{K}$ since the description is more transparent in this case. In this case there is an algebraic group defined and split over $\mathbb{Q}$ and in fact a group scheme over $\mathbb{Z}$ such that $\underline{G}$ is obtained from it by extension of scalars. We will abuse notation and also denote the above group-scheme over $\mathbb{Z}$ by $\underline{G}$ as well. Under this assumption the torus $\underline{T}$ is also defined and split over $\mathbb{Q}$, hence the group of real points of $\underline{T}$ is well-defined. We thus consider the real torus $\underline{T}(\mathbb{R})$ with the *real* Lie algebra $\mathfrak{a}$. The map Φ of Definition 2.6 (with $\mathbb{F} = \mathbb{R}$) gives us an identification

$$X_*(\underline{T}) \otimes \mathbb{R} \cong \mathfrak{a}.$$

Then the group $X_*(\underline{T})$ is identified with a lattice $L := L_{\underline{G}}$ in $\mathfrak{a}$. Similarly we use the map $\Phi^\vee$ to identify $X^*(\underline{T})$ with a lattice in the dual space $\mathfrak{a}^*$. The root system R sits naturally in the dual space $\mathfrak{a}^*$ and the corresponding finite Weyl group W_{sph} acts on $\mathfrak{a}$ and $\mathfrak{a}^*$ in the usual way; W_{sph} leaves invariant the lattices $X_*(\underline{T})$ and $X^*(\underline{T})$. Let $\Delta \subset E$ be the (positive) Weyl chamber of the group W_{sph}, determined by the parabolic subgroup $\underline{P}$. Our goal is to prove that the space $K\backslash G/K$ can be identified with the "cone"

$$\Delta_L = \Delta \cap L.$$

By definition, a cocharacter ϕ is a homomorphism from the group $\mathbb{G}_m(\mathbb{K})$ to the group T: note that the map $\Psi : X_*(\underline{T}) \to T$ given by $\Psi(\phi) = \phi(\pi)$ is injective (where π is a fixed uniformizer). We define the subgroup A by

$$A = \Psi(X_*(\underline{T})).$$

The following lemma is immediate (it suffices to prove it for $GL(1)$).

LEMMA 3.2. *The group T is a direct product $T = A \times B$.*

We let $A_{\Delta_L} \subset A$ be the image of $\Delta_L = X_*(\underline{T}) \cap \Delta$ under the map Ψ.

LEMMA 3.3. *(See* [**Tits**, page 51]*.) We have the Cartan decomposition*

$$G = K \cdot A_{\Delta_L} \cdot K,$$

i.e., each element $g \in G$ *has a unique representation as a product* $k_1 \cdot a \cdot k_2$*, where* $a \in A_{\Delta_L}$ *and* $k_i \in K$.

COROLLARY 3.4. *The inclusion* $\iota : A_{\Delta_L} \subset A \subset T \subset G$ *induces a bijection* $A_{\Delta_L} \xrightarrow{\iota} K\backslash G/K$.

Now that we have identified the double coset space

$$\Sigma_{inv\ fact}(G) = K\backslash G/K$$

with Δ_L, it is clear that the generalized invariant factor problem **Q3** in the Introduction agrees with the double coset problem $\mathbf{R}(G)$. However it may not be clear at this point why these problems agree with Problem **P3** in case $G = GL(\ell)$.

To see this we note that we can give a second description of the vector $\Phi(\lambda)$ if we choose coordinates in $\underline{T}$. Let $\chi_1, \cdots, \chi_n$ be a basis for the group $X^*(\underline{T})$ and define $\chi : T \longrightarrow \mathbb{K}^n$ to be the map with components $\chi_j, 1 \le j \le n$. We observe that we may use the derivative of χ_j at e to obtain coordinates on $T_e(\underline{T})$ and accordingly we obtain a map $\dot{\chi} : \mathfrak{a} \longrightarrow \mathbb{R}^n$. We leave the following lemma to the reader.

LEMMA 3.5. $\dot{\chi}(\Phi(\lambda)$ *is the (integral) vector in* $\mathbb{R}^n$ *obtained by applying the valuation* v *to the coordinates of* $\chi(\lambda(\pi))$.

General Case. We now consider the general case when $\underline{G}$ is not necessarily split. Most of the above discussion remains valid, however the Cartan decomposition for G has a slightly different form and for this reason we *cannot* use the cocharacter lattice as $L_{\underline{G}}$. We no longer can talk about *real points* of $\underline{T}$, so we do not have the interpretation of $\mathfrak{a}$ as a tangent space. We *define* $\mathfrak{a}$ as $X_*(\underline{T}) \otimes \mathbb{R}$. We set $V := \mathfrak{a}$.

Let E denote the affine space underlying V; as before, $\Delta \subset V$ is the positive Weyl chamber for W_{sph}. Note that $X^*(\underline{Z}) \subset X^*(\underline{T})$ is a subgroup of finite index.

DEFINITION 3.6. Define the homomorphism $\nu : Z \to V$ by the formula (see [**Tits**])

$$\langle \chi, \nu(z) \rangle = -v(\chi(z)), \forall z \in Z, \chi \in X^*(\underline{Z}). \tag{3.2}$$

Note that on the left hand side of the above formula, the pairing of $\chi \in X^*(\underline{Z}) \subset X^*(\underline{T})$ with $\nu(z)$ is coming from the pairing $\langle \cdot, \cdot \rangle$ between $X_*(\underline{T})$ and $X^*(\underline{T})$. On the right hand side of the equation (3.2), the action of χ on z is coming from the fact that each character $\chi \in X^*(\underline{Z})$ defines a character of the group $Z = \underline{Z}(\mathbb{K})$, i.e. we have $\chi(z) \in \mathbb{K}^\times$.

Let Z_c denote the kernel of ν and let $L_{\underline{G}}$ denote the image of ν in V. Then Z_c is a maximal compact subgroup of Z, see [**C**, pg. 135].

DEFINITION 3.7. We will refer to the group $L := L_{\underline{G}} := \nu(Z)$ as the *extended cocharacter lattice* of $\underline{T}$.

We have the inclusions

$$X_*(\underline{T}) = Hom(X^*(\underline{T}), \mathbb{Z}) \subset L_{\underline{G}} \subset Hom(X^*(\underline{Z}), \mathbb{Z})$$

with equality in the case when $\underline{G}$ is split over $\mathbb{Q}$. The quotient group N/Z_c operates on the affine space E faithfully, through a discrete group $\tilde{W}$. The finite Weyl group W_{sph} is the linear part of $\tilde{W}$ (the stabilizer of the origin) and we have $\tilde{W} = W_{sph} \ltimes L_{\underline{G}}$.

We refer to [**Tits**, section 1.2] for more details. Let $\Delta_L := \Delta \cap L$ and $Z_\Delta := \nu^{-1}(\Delta)$. Then one has:

THEOREM 3.1. [**Tits**, section 3.3.3] $G = KZK$ *and the map* $K\backslash G/K \to V$ *given by*

$$KzK \mapsto \nu(z), z \in Z_\Delta,$$

descends to a bijection $K\backslash G/K \to \Delta_L$.

This proves that in the nonsplit case we can also identify Δ_L with the parameter space $\Sigma_{inv\ fact}(G) = K\backslash G/K$ for the Problem **Q3**.

3.4. Comparison of the parameter spaces for the four algebra problems

We can now compare the four algebra problems **Q1**–**Q4** stated in the Introduction. We have already compared the parameter spaces for **Q3** and **Q4** in the Introduction. To compare the first three problems we assume that we are given groups $\underline{G}_1, \underline{G}_2, \underline{G}_3$ which are connected reductive algebraic groups (possibly all the same), where the first two are defined over $\mathbb{R}$, and $\underline{G}_3$ is over $\mathbb{F}_{(3)} = \mathbb{K}$, where $\mathbb{K}$ is a nonarchimedean valued field with valuation v and value group $\mathbb{Z}$. Suppose that the fields $\mathbb{F}_{(1)}, \mathbb{F}_{(2)}$ are either $\mathbb{R}$ or $\mathbb{C}$. We assume that all three Lie groups $G_i = \underline{G}_i(\mathbb{F}_{(i)})$ have the same (relative) rank l and isomorphic (relative) Weyl groups $W = W_{sph}$ acting on the appropriately chosen vector space $\mathfrak{a}$. Let $\Delta \subset \mathfrak{a}$ be a (positive) Weyl chamber. Let $L = L_{\underline{G}_3}$ be the extended cocharacter lattice, see Definition 3.7. Then the parameter spaces Σ_1, Σ_2 for the Problems **Q1** and **Q2** (for the groups G_1, G_2 resp.) are exactly the same: they are equal to Δ. The parameter space $\Sigma_{inv\ fact}(G_3)$ is the intersection Δ_L of the lattice L with Δ.

Note that in the special case when $\underline{G} = \underline{G}_i$, $i = 2, 3$, is split over $\mathbb{Q}$, $L = X_*(\underline{T})$ and we get a canonical isomorphism

$$X_*(\underline{T}) \otimes \mathbb{R} \to \mathfrak{a},$$

described in the previous section (here $\mathfrak{a}$ is the *real* Lie algebra of $\underline{T}(\mathbb{R})$). This isomorphism identifies $\Sigma_{inv\ fact}(G_3)$ with the set of "integer points" Δ_L in the Weyl chamber Δ.

3.5. Linear algebra problems

In this section we reformulate problems **Q1**– **Q3** stated in the introduction as *linear algebra* problems in the case when the group $\underline{G}$ is $GL(m)$.

Let α, β and γ be m-tuples of real numbers arranged in decreasing order. In Problem **P3** we let $\mathbb{K}$ be a complete, nonarchimedean valued field. We assume that the valuation v is discrete and takes values in $\mathbb{Z}$. We let $\mathcal{O} \subset \mathbb{K}$ be the subring of elements with nonnegative valuation.

In order to state Problems **P2** and **P3** below we recall some definitions from algebra. The *singular values* of a matrix M are the (positive) square-roots of the eigenvalues of the matrix MM^*. To define the *invariant factors* of a matrix M with entries in $\mathbb{K}$ note first that it is easy to see that the double coset

$$GL(m, \mathcal{O}) \cdot M \cdot GL(m, \mathcal{O}) \subset GL(m, \mathbb{K})$$

is represented by a diagonal matrix $D := D(M)$. The invariant factors of M are the integers obtained by applying the valuation v to the diagonal entries of D. If we arrange the invariant factors in decreasing order they are uniquely determined by M.

Now we can state the three algebra problems that interest us here, following the presentation in [**Fu1**].

- **P1. Eigenvalues of a sum.** Give necessary and sufficient conditions on α, β and γ in order that there exist Hermitian matrices A, B and C such that the set of eigenvalues (arranged in decreasing order) of A, resp. B, resp. C is α, resp. β, resp. γ and
$$A + B + C = 0.$$
- **P2. Singular values of a product.** Give necessary and sufficient conditions on α, β and γ in order that there exist matrices A, B and C in $GL(m, \mathbb{C})$ the logarithms of whose singular values are α, β and γ, respectively, so that
$$ABC = 1.$$
- **P3. Invariant factors of a product.** Give necessary and sufficient conditions on the integer vectors α, β and γ in order that there exist matrices A, B and C in $GL(m, \mathbb{K})$ with invariant factors α, β and γ, respectively, so that
$$ABC = 1.$$

We now see that the problems **Pi** are equivalent to the corresponding problems **Qi**; we consider the field $\mathbb{F} = \mathbb{C}$ since for $\mathbb{F} = \mathbb{R}$ the discussion is similar.

1. We have $\mathfrak{g} = \mathfrak{gl}(m)$ and the Cartan subspace $\mathfrak{p}$ is the space of Hermitian m-by-m matrices, $\mathfrak{a}$ is the space of diagonal m-by-m matrices with real entries and the cone $\Delta \subset \mathfrak{a}$ is the cone in which the diagonal entries are arranged in decreasing order. The maximal compact subgroup $K = U(m)$. Thus the parameter space Σ for the double coset $K\backslash GL(m, \mathbb{C})/K$ is the cone of m-tuples of real numbers arranged in decreasing order. This agrees with the parameter space of the Problem **P1**.

2. Observe that the Cartan subgroup $A \subset GL(m, \mathbb{C})$ consists of diagonal matrices with positive diagonal entries, A_Δ consists of matrices $D \in A$ with diagonal entries arranged in the decreasing order, and the projection f of g to $KgK \in \Sigma = A_\Delta \cong \Delta$ is given by first sending g to $h = \sqrt{gg^*}$, then diagonalizing h, arranging the diagonal values in the decreasing order and then taking their logarithms. Thus $f(g)$ is given by the vector whose components are the logarithms of the singular values of the matrix g. We conclude that Problem **P2** is equivalent to the Problem **Q2**.

3. The torus $T \subset G = GL(m, \mathbb{K})$ consists of diagonal matrices with entries in $\mathbb{K}$; we find that $\Psi(\phi) = \phi(\pi)$ is the diagonal matrix with diagonal entries $(\pi^{x_1}, \cdots, \pi^{x_m})$, where ϕ corresponds to the lattice vector $(x_1, \cdots, x_m)$. Therefore **P3** $\iff$ **Q3**.

Historic sketch.

To unify the notation, in what follows we will refer to the problem **Q4** for the group $GL(m)$ as **P4**.

The complete solution of Problems **P1**–**P4** and the relation between them were established only recently due to the efforts of several people. After much classical work, A. Klyachko in [**Kly1**] proved that Sol(**P1**) is a polyhedral cone in $\mathbb{R}^{3m}$, he also computed a finite system of linear homogeneous inequalities describing Sol(**P1**) in terms of the Schubert calculus in the Grassmannians $G(p, \mathbb{C}^m)$. This computation was subsequently improved by P. Belkale [**Bel**].

A. Klyachko then proved in [**Kly2**] that Sol(**P1**)=Sol(**P2**), which was known as *Thompson's conjecture.*

The sets of solutions for Problems **P3** and **P4** are also the same, namely they are integral points in the above polyhedral cone. This result is due to P. Hall, J. Green and T. Klein, see [**Kle1**] and [**Kle2**]; see also [**Mac**, pg. 100] for the history of this problem.

In this paper we show (Theorem 8.5) that the inclusion

$$Sol(\mathbf{Q4}) \subset Sol(\mathbf{Q3})$$

is true for *all* split reductive groups over $\mathbb{Q}$. This was the harder of the two implications for $GL(m)$, proved by T. Klein in [**Kle1**] and [**Kle2**], see also [**Mac**, pg. 94–100].

The exceptional (i.e. not true for all split reductive groups) inclusion

$$Sol(\mathbf{P}3) \subset Sol(\mathbf{P}4)$$

was first proved for $GL(m)$ by P. Hall but not published. In fact it follows from a beautiful and elementary observation of J. Green, which is set forth and proved in [**Mac**, pg. 91-92], and which we will outline in §8.6.

The description of the solutions to Problem **P4** as the set of integral points in the polyhedral cone Sol(**P1**) (known as the *Saturation Conjecture*) is due to A. Knutson and T. Tao, [**KT**] (see also [**DW**] for a proof using quivers and [**KM2**] for a proof in the spirit of the present paper using Littelmann's path model). The Knutson-Tao theorem combined with the above equivalence of Problems **P3** and **P4** establishes that the set of solutions to Problem **P3** is also the set of integral points in the above polyhedral cone.

In the present paper we reverse this path: We first prove directly that the set of solutions to Problem **P3** is the set of integral points in the above polyhedral cone, then using the above equivalence of Problems **P3** and **P4** we deduce that this set is also the set of solutions to Problem **P4**, thus providing yet another proof of the saturation conjecture .

We refer to [**AW**], [**AMW**], [**Bel**], [**BK**], [**BeSj**] [**Fu1**], [**EL**], [**KM2**], [**Kly1**], [**Kly2**] for more details and further developments.

CHAPTER 4

The existence of polygonal linkages and solutions to the algebra problems

In this chapter we will show that the algebra problems $\mathbf{R}(\overline{G})$ from chapter 3 can be restated geometrically, in terms of the existence of triangles with the prescribed "side-lengths" in three classes of spaces of nonpositive curvature, which are:

(1) X is an *infinitesimal symmetric space* (a Cartan motion space), $X = \mathfrak{p}$.
(2) X is a symmetric space of nonpositive curvature.
(3) X is an Euclidean building.

4.1. Setting up the general geometry problem

We start with a purely set-theoretic discussion. Let X be a set with a base-point o, and let $\overline{G}$ is a group acting on X. Let K denote the stabilizer of o in G. In general, the $\overline{G}$-action is not transitive (such examples will appear when X is a Euclidean building), we let Y denote the orbit $G \cdot o$. For pairs of points $(x, x') \in X^2$ we define the invariant

$$\sigma_{\bar{G}}(x, x')$$

as the projection of (x, x') to $\overline{G}\backslash X^2$. We will regard $\sigma_{\bar{G}}$ as the "generalized $\overline{G}$-invariant distance" between points in X. Given $\tau \in \overline{G}\backslash X^2$ we regard the set

$$S(o, \tau) = \{x \in X : \sigma_{\bar{G}}(o, x) = \tau\},$$

as the "sphere of radius τ" centered at o.

For pairs of points $(x, y) \in Y^2$ one can interpret the invariant $\sigma_{\bar{G}}(x, y)$ as follows: Let Σ denote the quotient space

$$\Sigma = K\backslash Y = K\backslash \overline{G}/K.$$

Then for pairs of points $(x, y) = (gK, hK)$ in Y we can identify $\sigma_{\bar{G}}(x, y)$ with the double coset $Kg^{-1}hK$. In other words, translate the pair (x, y) by g^{-1} to $(o, z) = (o, g^{-1}(o))$ and then project $g^{-1}(o)$ to the element $\sigma_{\bar{G}}(x, y) \in K\backslash Y$. Then two pairs (x, y) and (x', y') in Y^2 belong to the same $\overline{G}$-orbit if and only if

$$\sigma_{\bar{G}}(x, y) = \sigma_{\bar{G}}(x', y').$$

Hence, $\overline{G}\backslash Y \times Y$ can be bijectively identified with $\Sigma = K\backslash \overline{G}/K$.

We will use the notation Υ for the map $G \to \Sigma$. For $\sigma \in \Sigma$ we let $\mathcal{O}_\sigma \subset \overline{G}$ denote $\Upsilon^{-1}(\sigma)$.

It is now clear that the Problem $\mathbf{R}(\overline{G})$ stated in chapter 3 can be restated as a special case ($n = 3$) of the following:

PROBLEM 4.1. Give conditions on the vector of generalized side-lengths $\vec{\sigma}=(\sigma_1,\cdots,\sigma_n)\in\Sigma^n$ that are necessary and sufficient in order that there exist elements $g_1,g_2,\cdots,g_n$ in $\overline{G}$ such that

$$\Upsilon(g_i)=\sigma_i, 1\leq i\leq n, \quad \text{and} \quad \prod_{i=1}^{n} g_i = id.$$

We set $\mathcal{O}=\prod_{i=1}^{n}\mathcal{O}_{\sigma_i}$. Note that $(K\times K)^n$ acts on $\mathcal{O}$ by right and left multiplications.

Motivated by the connection between the moduli spaces of n-gons and symplectic quotients when Y is a complete simply-connected 3-dimensional Riemannian manifold of constant curvature (see [**KM1**], [**KMT**]), [**Tr**]) we define the "momentum map" $\mu:\mathcal{O}\longrightarrow\overline{G}$ by

$$\mu(g_1,...,g_n)=g_1\cdot g_2\cdots g_n.$$

We define the analogue of the symplectic quotient of $\mathcal{O}$ by K, by

$$\mathcal{O}//K=\{g\in\mathcal{O}:\mu(g)=id\}/K.$$

Here we divide out by the diagonal action of K, where K acts on each factor by *conjugation.* This action is the only action of K on $\mathcal{O}$ that we will use henceforth.

We let $\mathcal{Q}_\sigma(\overline{G})$ be the subset of $g\in\mathcal{O}$ such that $\mu(g)=id$. Thus $\mathcal{Q}_\sigma(\overline{G})$ is the set of solutions of the problem 4.1 and $\mathcal{O}//K$ is its quotient by K.

Our goal now is to reformulate the problem 4.1 in more geometric terms.

An *n-gon* (or *a closed n-gonal linkage*) in X with the vertices $x_1,...,x_n\in X$ is an n-tuple $\vec{x}=(x_1,...,x_n)\in X^n$, regarded as a map $\mathbb{Z}/n\mathbb{Z}\to X$. The *generalized side-lengths* of the polygon $\vec{x}$ are the elements

$$\sigma_i=\sigma_{\bar{G}}(x_i,x_{i+1})\in\overline{G}\backslash X^2, \quad i\in\mathbb{Z}/n\mathbb{Z}.$$

Thus for each n-gon we get a vector $\vec{\sigma}=(\sigma_1,...,\sigma_n)$ of its generalized side-lengths. The group $\overline{G}$ acts naturally on the space $\mathcal{P}ol_n(X)$ of all n-gons, preserving the generalized side-lengths of the polygons.

Fix $\vec{\sigma}=(\sigma_1,...,\sigma_n)$ and form the space

$$\mathcal{P}ol_{n,\sigma}(X):=\{P\in\mathcal{P}ol_n(X):\sigma_{\bar{G}}(x_i,x_{i+1})=\sigma_i\},$$

and its quotient, the *moduli space* of polygons with the fixed side-lengths in X:

$$\mathcal{M}_{n,\sigma}(X):=\mathcal{P}ol_{n,\sigma}(X)/\overline{G}. \tag{4.1}$$

In the case when X is a topological space and $\overline{G}$ acts homeomorphically on X we give the space $\mathcal{M}_{n,\sigma}(X)$ the quotient topology.

We note that $\mathcal{P}ol_n(Y)$ sits naturally in $\mathcal{P}ol_n(X)$ as the subset of polygons with vertices in Y. The generalized side-lengths of the polygons with vertices in Y are regarded as elements of $\Sigma=K\backslash Y$.

We define a map $\Phi:\mathcal{Q}_\sigma(\overline{G})\to Y^n$ by

$$\Phi(\vec{g})=(o,g_1\cdot o,g_1g_2\cdot o,\cdots,g_1g_2\cdots g_{n-1}\cdot o).$$

We see that Φ is K-equivariant (recall that K acts on $\mathcal{O}$ by conjugation on each factor).

LEMMA 4.2. *The map Φ induces a surjection $\overline{\Phi}$ from $\mathcal{O}//K$ onto the moduli space $\mathcal{M}_{n,\sigma}(Y)$.*

Proof: We first verify that each polygon $\Phi(\vec{g})$ has the "correct" side-lengths. By the left-invariance of $\sigma_{\bar{G}}$ we have

$$\sigma_{\bar{G}}(\prod_{i=1}^{k} g_i \cdot o, \prod_{i=1}^{k+1} g_i \cdot o) = \sigma_{\bar{G}}(o, g_k \cdot o) = \sigma_k.$$

To prove surjectivity of the map $\overline{\Phi}$ let $\vec{y} = (y_1, ..., y_n) \in \mathcal{P}ol_{n,\vec{\sigma}}(X)$. Choose g such that $g \cdot y_1 = o$ and replace $\vec{y}$ by $g^{-1} \cdot \vec{y}$. Since $\sigma_{\bar{G}}(y_1, y_2) = \sigma_1$ and (now) $y_1 = o$ we find that $y_2 \in S(o, \sigma_1)$. Hence there exists $g_1 \in \mathcal{O}_{\sigma_1}$ such that $y_2 = g_1 \cdot o$. Similarly, there exists $g_2 \in \mathcal{O}_{\sigma_2}$ such that $g_1 g_2 \cdot o = y_3$. Continuing in this way we get an n-tuple $(g_1, \ldots, g_{n-1}, g_n') \in \mathcal{O}$ such that

$$(o, g_1 \cdot o, g_1 g_2 \cdot o, \cdots, g_1 g_2 \cdots g_{n-1} \cdot o, g_1 g_2 \cdots g_{n-1} g_n' \cdot o) = (y_1, ..., y_n, y_1),$$

where $y_1 = o$. Thus $g_1 g_2 \cdots g_{n-1} g_n' = k \in K$ and after replacing g_n' with $g_n = g_n' k^{-1}$ we get an n-tuple $\vec{g} = (g_1, ..., g_n) \in \mathcal{O}$ such that $\mu(\vec{g}) = id$ and $\Phi(\vec{g}) = \vec{y}$. □

Thus we find that the problem 4.1 can be solved if and only the moduli space $\mathcal{M}_{n,\sigma}(Y)$ is nonempty.

To prove a version of Lemma 4.2 with a *bijective* map and to get a better analogy with the symplectic quotients we have to make a further assumption about the action $\overline{G} \curvearrowright Y$:

ASSUMPTION 4.3. Through the rest of this section we assume that $\overline{G}$ contains a subgroup B which acts simply-transitively on Y.

Now, instead of the orbits $\mathcal{O}_{\sigma_i}$ and their product $\mathcal{O}$ we consider the intersections:

$$\widehat{\mathcal{O}}_{\sigma_i} := \mathcal{O}_{\sigma_i} \cap B, \text{ and } \widehat{\mathcal{O}} := \mathcal{O} \cap B^n.$$

The group K no longer acts on $\widehat{\mathcal{O}}$ by conjugations, instead one has the use a *dressing action*, $dress(K)$, see [**KMT**] for the definition in the case when $X = Y = \mathbb{H}^3$, $G = PSL(2, \mathbb{C})$ and B fixes a point at infinity. In the case of infinitesimal symmetric spaces $X = \mathfrak{p}$ (where $\mathfrak{p} \oplus \mathfrak{k} = \mathfrak{g}$) one takes $B = \mathfrak{p}$ and the adjoint action of K as the dressing action. We now redefine the symplectic quotient as follows:

$$\widehat{\mathcal{O}}//K := (\widehat{\mathcal{O}} \cap \mu^{-1}(id))/dress(K).$$

Then one gets an analogue of Lemma 4.2 (which we do not need for the purposes of this paper), we refer the reader to [**KMT**] for the discussion in the case of $X = \mathbb{H}^3$:

LEMMA 4.4. *The map Φ induces a bijection $\widehat{\Phi}$ from $\widehat{\mathcal{O}}//K$ onto the moduli space $\mathcal{M}_{n,\sigma}(Y)$. Moreover, let Y be a topological space and $G \subset Homeo(Y)$. Then the quotient spaces $\widehat{\mathcal{O}}//K$ and $\mathcal{M}_{n,\sigma}(Y)$ have natural topology and the bijection $\widehat{\Phi}$ is a homeomorphism.*

The above discussion has been completely formal, our next goal is to describe the spaces X which can be used to analyze the problem 4.1 for various groups $\overline{G}$.

4.2. Geometries modeled on Coxeter complexes

Fix a spherical or Euclidean Coxeter complex (A, W), where A is a Euclidean space E or a sphere S and $W = W_{aff}$ or $W = W_{sph}$ is a (possibly nondiscrete) Euclidean or a spherical Coxeter group acting on A. If W is discrete then the Coxeter complex (A, W) is called *discrete*. In the case of Euclidean Coxeter complexes we let $L_{trans} \subset W$ denote the translation subgroup of W. Pick a special vertex $o \in E$ with stabilizer W_{sph} (see chapter 2) and let $\Delta \subset E$ be a Weyl chamber of W_{sph}. We will use the notation Δ_{sph} to denote the ideal boundary of Δ; Δ_{sph} is contained in the sphere S which will be regarded as the sphere at infinity of E; thus Δ_{sph} is a fundamental domain for the action $W_{sph} \curvearrowright S$.

Let Z be a metric space. A *geometric structure* on Z *modeled on* (A, W) consists of an atlas of isometric embeddings $\varphi : A \hookrightarrow Z$ satisfying the following compatibility condition: For any two charts φ_1 and φ_2, the transition map $\varphi_2^{-1} \circ \varphi_1$ is the restriction of an isometry in W. The charts and their images, $\varphi(A) = a \subset Z$, are called *apartments*. We will sometimes refer to A as the *model apartment*. We will require that there are *plenty of apartments* in the sense that any two points in Z lie in a common apartment. All W-invariant notions introduced for the Coxeter complex (A, W), such as walls, singular subspaces, chambers etc., carry over to geometries modeled on (A, W).

One defines the group of automorphisms $Aut(A)$ of the model apartment as the group of isometries of A which normalize the subgroup W. If X is a space modeled on (A, W) then an isometry $g : X \to X$ is an automorphism if it sends apartments to apartments and for each pair of apartments $(A, \varphi_1), (A, \varphi_2)$ the composition $\varphi_2^{-1} \circ g \circ \varphi_1$ is the restriction of an automorphism of A. The group of automorphisms of X is denoted $Aut(X)$.

Examples of the above geometries are provided by symmetric spaces of noncompact type and their infinitesimal analogues (infinitesimal symmetric spaces). These are modeled on Euclidean Coxeter complexes with transitive affine Weyl group. In the case of a symmetric space X, the apartments are the maximal flats. The associated Coxeter complex has the form (E, W_{aff}) where E is an apartment and W_{aff} is the group generated by reflections at singular hyperplanes.

Take a real or complex reductive Lie group G, the Lie algebra $\mathfrak{g}$ of G has the decomposition $\mathfrak{g} = \mathfrak{p} \oplus \mathfrak{k}$, where $K \subset G$ is a maximal compact subgroup, let $\mathfrak{k}$ denote the Lie algebra of K. We will identify the Cartan subspace $\mathfrak{p}$ with the tangent space T_pX to the symmetric space $X = G/K$ at the point p stabilized by K. The subspaces $\mathfrak{p} \subset \mathfrak{g}$, equipped with the affine action of $\overline{G} = \mathfrak{p} \rtimes K$, are *infinitesimal symmetric spaces* in the following sense. The (E, W_{aff})-structure on X induces a (E, W_{aff})-structure on the *Cartan subspace* $\mathfrak{p} = T_pX$, such that the apartments in $\mathfrak{p}$ are the translates of the Cartan subalgebras (i.e. the maximal abelian subalgebras) in $\mathfrak{p}$. The apartments through 0 in $\mathfrak{p}$ are the tangent spaces to the apartments through p in X. The term "infinitesimal symmetric space" may be also justified by noting that there is a one-parameter family of spaces X_ϵ parameterized by $\epsilon \geq 0$ and all isometric to X for $\epsilon > 0$ such that $X_0 = \mathfrak{p}$, see [**KMT**, §5].

The last kinds of geometries considered in this paper are spherical and Euclidean buildings. We refer the reader to [**Ba**] for the definitions of $CAT(\kappa)$ metric spaces.

DEFINITION 4.5. A *spherical building* is a CAT(1)-space modeled on a spherical Coxeter complex.

We will use the notation $\angle_{Tits}$ for the metric in a spherical building. Spherical buildings have a natural structure as polysimplicial piecewise spherical complexes. We prefer the geometric to the combinatorial view point because it appears to be more flexible.

DEFINITION 4.6. A discrete Euclidean building is a CAT(0)-space modeled on a discrete Euclidean Coxeter complex.

In the non-discrete case the definition of Euclidean buildings is more subtle, see [**KL**, section 4.1.2]. We refer to [**KL**] for a thorough discussion of buildings from the geometric viewpoint.

A building is called *thick* if every wall is an intersection of apartments. A non-thick building can always be equipped with a natural structure of a thick building by reducing the Weyl group.

EXAMPLE 4.7. If X is a symmetric space of noncompact type or a thick Euclidean building modeled on the Coxeter complex (E, W_{aff}), then its ideal boundary $\partial_{Tits}X$ is a thick spherical building modeled on $(\partial_{Tits}E, W_{sph})$. In the case that X is a building, the *spaces of directions* $\Sigma_x X$ are spherical buildings modeled on $(\partial_{Tits}E, W_{sph})$. The building $\Sigma_x X$ is thick if and only if x is a special vertex of X. We note that in the case when X is a discrete building, $\Sigma_x X$ is just the link of the point $x \in X$.

Let B be a spherical building modeled on a spherical Coxeter complex (S, W_{sph}). We say that two points $x, y \in B$ are antipodal, if $\angle_{Tits}(x, y) = \pi$; equivalently, they are antipodal points in an apartment $s \subset B$ containing both x and y. The quotient map $S \to S/W_{sph} \cong \Delta_{sph}$ induces a canonical projection $\theta : B \to \Delta_{sph}$ folding the building onto its model Weyl chamber. The θ-image of a point in B is called its *type*.

REMARK 4.8. To define $\theta(x)$ pick an apartment s containing x and a chart $\phi : S \to s$. Then $\theta(x)$ is the projection of $\phi^{-1}(x)$ to $S/W_{sph} \cong \Delta_{sph}$. We note that this is clearly independent of s and ϕ.

The same definition applies in the case of Euclidean buildings B. The difference however is that the action $W_{aff} \curvearrowright E$ in general is no longer discrete, so we cannot identify the image of $B \to E/W_{aff}$ with a simplex. If W_{aff} acts as a lattice on E, then E/W_{aff} can be identified with a *fundamental alcove* for the action $W_{aff} \curvearrowright E$.

We now give two properties of the projection θ:

1. If $h : a \to a'$ is an isomorphism of apartments in B (i.e. ${\phi'}^{-1} \circ h \circ \phi \in W$) then $\theta \circ h = \theta$.

2. If B is a spherical building, $x, x' \in B$ which belong to apartments a, a' respectively and $-x \in a, -x' \in a'$ are antipodal to x, x', then $\theta(x) = \theta(x')$ implies $\theta(-x) = \theta(-x')$. To prove this pick an isomorphism $h : a \to a'$. Then (since $\theta(x) = \theta(x')$) there exists $w \in W \curvearrowright a'$ such that $w(h(x)) = x'$. Hence $w \circ h(-x) = -w \circ (x) = -x'$. The claim now follows from 1.

DeRham decomposition of Euclidean buildings. Suppose that X is a thick Euclidean building which is modeled on a *reducible* discrete Coxeter complex (E, W_{aff}). Consider the deRham decomposition of the building X:

$$X = X_0 \times X_1 \times ... \times X_s,$$

where each X_i is a thick Euclidean building modeled on the Euclidean Coxeter complex (E_i, W^i_{aff}) and $X_0 \cong E_0$ is the flat deRham factor. For each $i = 1, ..., s$ the group W^i_{aff} acts as an irreducible affine Coxeter group on the Euclidean space E_i, for $i = 0$ we get the trivial Coxeter group W^0_{aff}. Accordingly, the Euclidean space E splits (metrically) as the product

$$E_0 \times \prod_{i=1}^{s} E_i,$$

this decomposition is invariant under W_{aff} which in turn splits as

$$W_{aff} = \prod_{i=0}^{s} W^i_{aff},$$

as explained in Section 2.

4.3. Bruhat-Tits buildings associated with nonarchimedean reductive Lie groups

In this section we describe properties of the Euclidean building (the *Bruhat-Tits' building*) associated to a reductive nonarchimedean Lie group by Bruhat and Tits. The reader familiar with [**BT**] can skip this section.

Let $\mathbb{K}$ be a valued field with valuation v and value group $\mathbb{Z}$. Let $\mathcal{O}$ denote the subring in $\mathbb{K}$ which consists of elements with nonnegative valuation. Let $\underline{G}$ be a connected reductive algebraic group over $\mathbb{K}$ which has relative rank l. The subgroup $K := \underline{G}(\mathcal{O}) \subset G = \underline{G}(\mathbb{K})$ is a maximal bounded subgroup of G. In this section we review the properties of a Euclidean building $X = X_{\underline{G}}$ attached to the group $\underline{G}$. We refer the reader to [**Tits**] and [**BT**, Chapter 7] for more details.

We begin by recalling the notation from section 3.3. Let $\underline{T} \subset \underline{G}$ be a maximal $\mathbb{K}$-split torus, $\underline{Z} \subset \underline{G}$ and $\underline{N} \subset \underline{G}$ be its centralizer and normalizer (over $\mathbb{K}$) respectively. We also get the groups N, Z and T of $\mathbb{K}$-points of the corresponding algebraic groups. Then $V := X_*(\underline{T}) \otimes \mathbb{R}$, where $X_*(\underline{T})$ is the group of cocharacters of $\underline{T}$. The Euclidean space E is the affine space underlying V with appropriately chosen Euclidean metric. Let $R_{rel} \subset V^*$ denote the relative root system of the pair $(\underline{G}, \underline{T})$; then W_{sph} is the finite Coxeter group corresponding to this system.

Definition 3.6 gives us a homomorphism $\nu : Z \to V$, with the kernel $Z_c \subset Z$ and the image equal to the extended cocharacter lattice $L_{\underline{G}}$. We also get the quotient group N/Z_c, which acts on the Euclidean space E discretely and isometrically through a group $\tilde{W}$. Unless $\underline{G}$ is semisimple, $\tilde{W}$ *is not* an affine Weyl group.

In the case when the group $\underline{G}$ is simply-connected and semisimple, one can take the pair $(E, \tilde{W})$ as the Euclidean Coxeter complex (E, W_{aff}) of the Bruhat-Tits building $X_{\underline{G}}$ attached to the group $\underline{G}$. In general however it is not the case and one has to do more work to define W_{aff}. The construction of this group will be unimportant for us (we refer to [**Tits**] for the explicit construction), the important properties of W_{aff} are the following:

(1) $\dim(E) = l$, the relative rank of $\underline{G}$ or the dimension of $\underline{T}$.
(2) $W_{aff} \subset \tilde{W}$ is a normal subgroup [**Tits**, section 1.7, page 34], the index $|\tilde{W} : W_{aff}|$ is finite in the semisimple case.
(3) $\tilde{W} = W_{sph} \ltimes L_{\underline{G}}$.
(4) W_{aff} is an affine Coxeter group attached to a root system $R \subset V^*$.
(5) The root system R in general *is not the same* as R_{rel} but they have the same rank and the same finite Weyl group W_{sph}, [**Tits**, section 1.7].
(6) $W_{aff} = W_{sph} \ltimes L_{trans}$, where W_{sph} is the finite Weyl group as above.

(7) We have the inclusions

$$Q(R^\vee) = L_{trans} \subset L_{\underline{G}} \subset N_{aff} = P(R^\vee).$$

(8) If $\underline{G}$ is a simply-connected semisimple group then $W_{aff} = \tilde{W}$, [**Tits**, section 1.13]. If $\underline{G}$ is split over an unramified extension of $\mathbb{K}$ then $L_{\underline{G}} = X_*(\underline{T})$, [**Tits**, section 1.3].

(9) If $\underline{G}$ is split then the reduced (Bruhat-Tits) root system associated with W_{aff} is the usual reduced root system R_{rel} of the group $\underline{G}$ (note that R_{rel} is also the absolute root system since we are in the split case).

Here as usual, L_{trans} is the translation subgroup of W_{aff} and N_{aff} is the normalizer of W_{aff} in V. When we are dealing with *the root system associated with* $\underline{G}$, as in 4 or 5 above, we will refer to R, and not to R_{rel}.

DEFINITION 4.9. We will call R the *Bruhat-Tits root system* associated with the algebraic group $\underline{G}$.

We refer the reader to Tableau des échelonages in [**BT**, p. 29-30] for computation of the Bruhat-Tits root systems.

We note that the root systems R and R_{rel} are not very different since they have the same finite Coxeter group. Thus, in the irreducible case, both root systems are either isomorphic or one of them is of type B_l and the other of type C_l. In particular, the *saturation factors* for R and R_{rel} are exactly the same (see chapter 7.1).

Having described the properties of the Coxeter complex (E, W_{aff}), we will describe the properties of the *Bruhat-Tits building* $X = X_{\underline{G}}$ associated with the group $\underline{G}$:

LIST 4.10. (1) *X is modeled on the Euclidean Coxeter complex (E, W_{aff}) described above.*

(2) *The group G acts on X by (isometric) automorphisms.*

(3) *The subgroup K is the stabilizer of a special vertex o in X.*

(4) *G acts transitively on the set of apartments in X.*

(5) *For each apartment $a \subset X$ let G_a be the stabilizer of a in G. Then the image of G_a in $Aut(a)$ is the group $\tilde{W}$ (containing W_{aff} as a normal subgroup).*

(6) *The building X is thick (see* [**BT**, Prop. 7.4.5]*).*

(7) *The group W_{aff} acts discretely and (in the semisimple case) cocompactly on E.*

REMARK 4.11. The properties in the List 4.10, except the last one, hold in the case of symmetric spaces and their infinitesimal analogues.

We note that the affine space E has the distinguished point o, the origin (corresponding to the trivial cocharacter). The corresponding vertex in X is also denoted by o, it is stabilized by $K = \underline{G}(\mathcal{O})$.

4.4. Geodesic polygons

In this paper we will be considering polygons in the metric spaces modeled on the Coxeter complexes, which were discussed in section 4.2. With the exception of the rank zero spherical buildings, all such metric spaces X are *geodesic* and thus we define a *geodesic polygon* in X as a polygon with the vertices $z_1 \dots z_n$ together with a choice of *sides* $\overline{z_i z_{i+1}}$, i.e. the geodesic segments connecting z_i to z_{i+1}. We note that in the case of metric spaces modeled on Euclidean Coxeter complexes the sides are uniquely determined by the polygon $z_1 \ldots z_n$. We recall that *n-gons* in X are regarded as maps $\mathbb{Z}/n\mathbb{Z} \to X$.

Let (A, W) be a spherical or Euclidean Coxeter complex. The complete invariant of a pair of points $(x, y) \in A^2$ with respect to the action $W \curvearrowright A$, is its image $\sigma_W(x, y)$ under the canonical projection to $A \times A/W$. We define the *refined length* of a geodesic segment $\overline{xy}$ as $\sigma_{ref}(x, y) := \sigma_W(x, y)$. This notion carries over to geometries modeled on the Coxeter complex (A, W): For a pair of points (x, y) pick an apartment a containing x, y and, after identifying a with the model apartment A, let $\sigma_{ref}(x, y)$ be the projection to $A \times A/W$.

In the case of Euclidean Coxeter complexes there are extra structures associated with the concept of refined length. Given a Euclidean Coxeter complex (E, W_{aff}), pick a special vertex $o \in E$. Then we can regard E as a vector space V, with the origin $0 = o$. Let $\Delta \subset E$ denote a Weyl chamber of W_{sph}, the tip of Δ is at o.

Suppose that L is a subgroup of the group V of all translations of E so that:

$$L_{trans} \subset L \subset N_{aff}, \tag{4.2}$$

where $N_{aff} \subset V$ is the normalizer of W_{aff}. Since o is the origin in E, we will identify the orbits $L \cdot o$ and $L_{trans} \cdot o$ with L and L_{trans} respectively. We define the set of *refined L-integral lengths* as the subset

$$(L \times L)/W_{aff} \subset (E \times E)/W_{aff}.$$

If $L = L_{trans}$ then we have a natural bijection

$$(L_{trans} \times L_{trans})/W_{aff} \cong \Delta \cap L_{trans}.$$

Since (E, W_{aff}) is a Euclidean Coxeter complex, there is also a coarser notion of *Δ-length* obtained from composing σ_{ref} with the natural forgetful map

$$E \times E/W_{aff} \to E/W_{sph} \cong \Delta.$$

To compute the Δ-length $\sigma(x, y)$ we regard the oriented geodesic segment $\overline{xy}$ as a vector in E and project it to Δ.

Again, the concepts of Δ-length, L-integral Δ-lengths, etc., carry over to the geometries modeled on (E, W_{aff}). Note that Δ-length and refined length coincide for symmetric spaces and their infinitesimal analogues because the affine Weyl group acts transitively. We define the set of *L-integral* Δ-lengths as the subset $\Delta \cap L \subset \Delta$. A segment with L-integral Δ-length has L-integral refined length iff its vertices lie in the distinguished orbit $L \cdot o$ (identified with L). For segments with endpoints of type $W_{aff} \cdot o$, the notions of Δ-length and refined length are equivalent.

Given a collection $\tau = (\tau_1, ..., \tau_n) \in \Delta^n$ of Δ-lengths we define the *moduli space* $\mathcal{M}_{n,\tau}(X)$ as the quotient of the collection of geodesic polygons in X with the Δ-side-lengths τ by the action of the group G. We give $\mathcal{M}_{n,\tau}(X)$ the quotient topology.

We are now ready to state the questions which will be (for $n = 3$) the geometric counterparts to the algebra questions in chapter 3:

PROBLEM 4.12. Let X be a symmetric space of noncompact type or an infinitesimal symmetric space, or a thick Euclidean building. Describe the set $D_n(X) \subset \Delta^n$ of Δ-side lengths which occur for geodesic n-gons in X.

REMARK 4.13. The above problem was solved in the papers [**KLM1, KLM2**]. In these papers the notation $\mathcal{P}_n$ was used in place of D_n. We refer the reader to section 5.2 for the description of the solution of Problem 4.12.

PROBLEM 4.14. Let X be a thick Euclidean building. Describe the set

$$D_n^{ref,L}(X) \subset (W\backslash L \times L)^n \subset (W\backslash A \times A)^n$$

of refined L-integral side-lengths which occur for n-gons in X.

One of the key results concerning the above problem is the following

THEOREM 4.1 (Transfer Theorem, see [**KLM2**]). *Suppose that X, X' are thick Euclidean buildings modeled on the Coxeter complexes $(A, W), (A', W')$ respectively, L, L' are lattices as in (4.2), $\iota : (A, W) \to (A', W')$ is an embedding of Coxeter complexes such that $\iota(L) \subset L'$. Then ι induces an embedding*

$$\iota_* : D_n^{ref,L}(X) \to D_n^{ref,L'}(X')$$

Below we explain how given an algebraic problem **Q1**–**Q3** one finds a metric space X modeled on a Euclidean Coxeter complex, so that the geometric problem 4.14 (for $n = 3$) is equivalent to the corresponding algebraic problem.

As we explained in the beginning of chapter 3, with each problem **Qi**, **i=1, 2, 3**, we can associate a pair of groups $K \subset \overline{G}$.

1. For the Problem **Q1** we take $\overline{G} = K \ltimes \mathfrak{p}$ and let $X := \mathfrak{p}$. Then X is a metric space modeled on the Euclidean Coxeter complex $(A, W) = (E, W_{aff})$.

2. For the Problem **Q2** we take $\overline{G} = G$ and let X be the symmetric space $X = G/K$.

In both cases $\overline{G}$ acts transitively on X and we apply Lemma 4.2 to the transitive action $\overline{G} \curvearrowright X$, to conclude that **Q1** and **Q2** are equivalent to Problem 4.12 for the corresponding space X.

EXAMPLE 4.15. Let $G = GL_m(\mathbb{C})$, $K = U(m)$. Then the symmetric space X associated with G is G/K, which is the space $\mathbb{P}_m$ of positive-definite Hermitian $m \times m$ matrices. The model apartment A in $\mathbb{P}_m$ consists of diagonal matrices with positive diagonal entries. The finite Weyl $W_{sph} = \mathcal{S}_m$ group acts on A by permutations of the diagonal elements; the affine Weyl group W_{aff} is isomorphic to $W_{sph} \ltimes \mathbb{R}^{m-1}$. Thus we get a geometric model for analyzing the Problem **P2** from the Introduction.

3. For the Problem **Q3** we assign to the group $G = \underline{G}(\mathbb{K})$ the Euclidean (Bruhat-Tits) building X as it was explained in the previous section. We let Y be the G-orbit of the special vertex $o \in X$.

Although, unlike in the previous two examples, the group G does not act transitively on the building X, this group acts transitively on the subset $Y \subset X$ and we apply Lemma 4.2 to the action $G \curvearrowright Y$ to see that the problem **Q3** (or, equivalently, $\mathbf{R}(G)$) is equivalent to finding necessary and sufficient conditions for existence of geodesic triangles in X whose vertices are in Y and whose Δ-lengths are the prescribed elements of Δ_L, where $L = L_{\underline{G}}$. More generally, there is a surjective map

$$pr : W\backslash A \times A \to G\backslash X \times X,$$

and by applying the Transfer Theorem 4.1 we get:

PROPOSITION 4.16. *Suppose that there exists a polygon P in X whose σ_G-side-lengths are $(\sigma_1, ..., \sigma_n)$. Then for any choice of $\tilde{\sigma}_i \in pr^{-1}(\sigma_i)$, there exists a polygon $\tilde{P}$ in X whose σ_{ref}-side-lengths are $\tilde{\sigma}_1, ..., \tilde{\sigma}_n$.*

Therefore we will stick to the notion of the refined side-length σ_{ref} through the rest of the paper.

CHAPTER 5

Weighted Configurations, Stability and the Relation to Polygons

Let X be a symmetric space of nonpositive curvature or a Euclidean building. Recall that the ideal boundary $B = \partial_{Tits}X$ has the structure of a spherical building, the metric on B is denoted by $\angle_{Tits}$. Given a Weyl chamber Δ in X, we get a spherical Weyl chamber $\Delta_{sph} = \partial_\infty \Delta \subset \partial_{Tits}X$. We will identify Δ_{sph} with the unit vectors in Δ. Recall that there is a canonical projection $\theta : \partial_{Tits}X \to \Delta_{sph}$, see section 4.2.

Take a collection of weights $m_1, ..., m_n \geq 0$ and define a finite measure space $(\mathbb{Z}/n\mathbb{Z}, \nu)$ where the measure ν on $\mathbb{Z}/n\mathbb{Z}$ is given by $\nu(i) = m_i$. An n-tuple of ideal points $(\xi_1, ..., \xi_n) \in B^n$ together with $(\mathbb{Z}/n\mathbb{Z}, \nu)$ determine a *weighted configuration at infinity*, which is a map

$$\psi : (\mathbb{Z}/n\mathbb{Z}, \nu) \to \partial_{Tits}X.$$

The *type* $\tau(\psi) = (\tau_1, \ldots, \tau_n) \in \Delta^n$ of the weighted configuration ψ is given by $\tau_i = m_i \cdot \theta(\xi_i)$. Let $\mu = \psi_*(\nu)$ be the pushed forward measure on B. We define the *slope* of a measure μ on B with finite total mass $|\mu|$ as

$$slope_\mu(\eta) = -\int_B \cos \angle_{Tits}(\xi, \eta)\, d\mu(\xi).$$

In this paper we will consider only measures with finite support.

DEFINITION 5.1 (Stability). A measure μ on B (with finite support) is called *semistable* if $slope_\mu(\eta) \geq 0$ and *stable* if $slope_\mu(\eta) > 0$ for all $\eta \in B$.

There is a refinement of the notion of semistability motivated by the corresponding concept in geometric invariant theory.

DEFINITION 5.2 (Nice semistability). A measure μ on B (with finite support) is called *nice semistable* if μ is semistable and $\{slope_\mu = 0\}$ is a subbuilding or empty. In particular, stable measures are nice semistable.

A weighted configuration ψ on B is called *stable, semistable* or *nice semistable*, respectively, if the corresponding measure $\psi_*\nu$ has this property.

For the purposes of this paper, i.e. the study of polygons, nice semistability plays a role in the case of symmetric spaces and infinitesimal symmetric spaces only. We note however that for these spaces, existence of a semistable configuration ψ on $\partial_{Tits}X$ implies existence of a nice semistable configuration on $\partial_{Tits}X$, which has the same type as ψ, see [**KLM1**].

EXAMPLE 5.3. (i) Let B be a spherical building of rank 0. Then a measure μ on B is stable iff it contains no atoms of mass $\geq \frac{1}{2}|\mu|$, semistable iff it contains no atoms of mass $> \frac{1}{2}|\mu|$, and nice semistable iff it is either stable or consists of two atoms of equal mass.

(ii) Suppose that B is a unit sphere and regard it as the ideal boundary of a Euclidean space E, $B = \partial_{Tits}E$. Each semistable measure μ has slope zero everywhere.

Define the subset $\Delta^n_{ss}(B) \subset \Delta^n$ consisting of those n-tuples $\tau \in \Delta^n$ for which there exists a weighted semistable configuration on B of type τ.

Suppose now that G is a reductive complex Lie group, $K \subset G$ is a maximal compact subgroup, $X = G/K$ is the associated symmetric space. Then the spaces of weighted configurations in $\partial_{Tits}X$ of the given type $\tau \in \Delta^n$ can be identified with products

$$F = F_1 \times ... \times F_n$$

where F_i's are smooth complex algebraic varieties (generalized flag varieties) on which the group G acts transitively. Hence G acts on F diagonally.

In case X is the symmetric space associated to a complex Lie group, the notions of stability (semistability, etc.) introduced above coincide with corresponding notions from symplectic geometry, and, in the case where the weights τ_i's are L-integral (i.e., belong to $L = L_{\underline{G}}$) they also coincide with the concepts of stability (semistability, etc.) used in the Geometric Invariant Theory; for a proof of this see [**KLM1**].

5.1. Gauss maps and associated dynamical systems

We now relate polygons in X (where X is a metric space modeled on a Euclidean Coxeter complex) and weighted configurations on the ideal boundary B of X, which plays a key role in [**KLM1**] and [**KLM2**]. If X is an infinitesimal symmetric space $\mathfrak{p}$ we identify the visual boundary of X with the Tits boundary $\partial_{Tits}X'$ of the corresponding symmetric space X' via the exponential map $\mathfrak{p} \to X'$. Thus for all three geometries, the ideal boundary B is a spherical building.

Consider a (closed) polygon $P = x_1x_2 \dots x_n$ in X, i.e. a map $\mathbb{Z}/n\mathbb{Z} \to X$. The distances $m_i = d(x_i, x_{i+1})$ determine a finite measure ν on $\mathbb{Z}/n\mathbb{Z}$ by $\nu(i) = m_i$. The polygon P gives rise to a collection $Gauss(P)$ of *Gauss maps*

$$\psi : \mathbb{Z}/n\mathbb{Z} \longrightarrow \partial_{Tits}X \tag{5.1}$$

by assigning to i an ideal point $\xi_i \in \partial_{Tits}X$ so that the geodesic ray $\overline{x_i\xi_i}$ (originating at x_i and asymptotic to ξ_i) passes through x_{i+1}. This construction, in the case of the hyperbolic plane, already appears in the letter of Gauss to W. Bolyai, [**G**]. Taking into account the measure ν, we view the maps $\psi : (\mathbb{Z}/n\mathbb{Z}, \nu) \to \partial_{Tits}X$ as *weighted configurations* of points on $\partial_{Tits}X$. Note that if X is a Riemannian symmetric space and the m_i's are all non-zero, there is a unique Gauss map due to the unique extendability of geodesics. On the other hand, if X is a Euclidean building then, due to the branching of geodesics, there are in general infinitely many Gauss maps. However, the corresponding weighted configurations are of the same type, i.e. they project to the same weighted configuration on Δ_{sph}.

The following crucial observation explains why the notion of semistability is important for studying closed polygons.

LEMMA 5.4 ([**KLM1**], [**KLM2**]). *For each Gauss map ψ the pushed forward measure $\mu = \psi_*\nu$ is semistable. If X is a symmetric space or an infinitesimal symmetric space then the measure μ is nice semistable.*

We are now interested in finding polygons with prescribed Gauss map. Such polygons will correspond to the fixed points of a certain dynamical system. For $\xi \in \partial_{Tits}X$ and $t \geq 0$, we define the map $\phi := \phi_{\xi,t} : X \to X$ by sending x to the point at distance t from x on the geodesic ray $\overline{x\xi}$. Since X is nonpositively curved, the map ϕ is 1-Lipschitz. Fix now a weighted configuration $\psi : (\mathbb{Z}/n\mathbb{Z}, \nu) \to \partial_{Tits}X$ with non-zero total mass. We define the 1-Lipschitz self-map

$$\Phi = \Phi_\psi : X \longrightarrow X$$

as the composition $\Phi_n \circ \cdots \circ \Phi_1$ of the maps $\Phi_i = \phi_{\xi_i, m_i}$. The fixed points of Φ are the first vertices of closed polygons $P = x_1 \dots x_n$ so that ψ is a Gauss map for P. Since the map Φ is 1-Lipschitz, and the space in question is complete and has nonpositive curvature, the existence of a fixed point for Φ reduces (see [**KLM2**]) to the existence of a bounded orbit for the dynamical system $(\Phi^n)_{n\in\mathbb{N}}$ formed by the iterations of Φ. Of course, in general, there is no reason to expect that $(\Phi^n)_{n\in\mathbb{N}}$ has a bounded orbit: for instance, if the support of the measure $\mu = \psi_*(\nu)$ is a single point, all orbits are unbounded.

One of our results is that under the appropriate semi-stability assumption on ψ the system $(\Phi^n)_{n\in\mathbb{N}}$ has a bounded orbit:

THEOREM 5.1 ([**KLM1**], [**KLM2**]). *Suppose that X is either a symmetric space or a Euclidean building with one vertex. Suppose that ψ is a nice semistable weighted configuration on $\partial_{Tits}X$ (in the symmetric space case) or a semistable configuration (in the building case). Then Φ_ψ has a fixed point.*

This theorem also holds for arbitrary Euclidean buildings: It was proven in an early version of [**KLM2**] under the local compactness assumption, this proof was superseded by a proof by Andreas Balser [**B**] who had removed the local compactness assumption.

Combining the above result with the Transfer Theorem 4.1 we get:

THEOREM 5.2 ([**KLM1, KLM2**]). *Suppose that X is a symmetric space of nonpositive curvature or an infinitesimal symmetric space or a Euclidean building with a model Weyl chamber Δ. Then $D_n(X) = \Delta^n_{ss}(\partial_{Tits}X)$.*

The equivalence of the Problems **Q1** and **Q2** (and consequently **P1** and **P2**) in the Introduction follows immediately from the above theorem since for an infinitesimal symmetric space $\mathfrak{p}$ and the corresponding symmetric space X the Tits boundaries are the same. As another corollary of the combination of Theorems 4.1 and 5.2 we get:

THEOREM 5.3 ([**KLM1, KLM2**]). *Let X be either a thick Euclidean building, a symmetric space or an infinitesimal symmetric space. Then $D_n(X)$ depends only on the associated spherical Coxeter complex and not on the type of the geometry.*

For instance, suppose that X is a nonpositively curved symmetric space and X' is a Euclidean building which have isomorphic finite Weyl groups and the same rank. Then

$$D_n(Cone(\partial_{Tits}X)) = \Delta^n_{ss}(\partial_{Tits}X) = D_n(X),$$
$$D_n(Cone(\partial_{Tits}X')) = \Delta^n_{ss}(\partial_{Tits}X') = D_n(X'),$$

where $Cone(\partial_{Tits}X)$ and $Cone(\partial_{Tits}X')$ are Euclidean cones over the Tits boundary of X, X', i.e. 1-vertex buildings. Since $Cone(\partial_{Tits}X)$ and $Cone(\partial_{Tits}X')$ have isomorphic spherical Weyl groups, their affine Weyl groups are isomorphic as well, so the transfer theorem implies that

$$D_n(Cone(\partial_{Tits}X)) = D_n(Cone(\partial_{Tits}X'))$$

as required.

In the case when the group G is complex, for the ideal boundaries B of symmetric spaces $X = G/K$ one constructs the moduli space of semistable weighted configurations on B as follows.

Given a type τ (so that for each i the vector τ_i is nonzero), the set of semistable configurations $Conf_{\tau,sst}(B)$ (resp. nice semistable configurations $Conf_{\tau,nsst}(B)$) of type τ on B has a natural topological structure. Define a relation $\sim$ on $Conf_{\tau,sst}(B)$ by $\psi \sim \psi'$ if

$$\overline{G \cdot \psi} \cap \overline{G \cdot \psi'} \neq \emptyset.$$

One then verifies that $\sim$ is an equivalence relation, see [**HL**] and [**Sj**]. Define the moduli space $\mathcal{M}_\tau(B)$ of semi-stable configurations on B of type τ as the quotient $Conf_{\tau,sst}(B)/\sim$. We note that the moduli space (see (4.1)) $\mathcal{M}_{n,\tau}(X)$ of n-gons in X with the given side-lengths $\tau \in \Delta^n$, is also a compact topological space.

THEOREM 5.4. *The moduli space $\mathcal{M}_{\tau,sst}(B)$ is Hausdorff. The map $P \mapsto Gauss(P)$ defines a natural homeomorphism $h : \mathcal{M}_{n,\tau}(X) \to \mathcal{M}_{\tau,sst}(B)$.*

Proof: It was proven in [**HL**] and [**Sj**] that $\mathcal{M}_{\tau,sst}(B) = Conf_{\tau,nsst}(B)/G$, and that $\mathcal{M}_{\tau,sst}(B)$ is Hausdorff. The surjection h is clearly continuous, it is also easily seen to be injective. Therefore the map

$$h : \mathcal{M}_{n,\tau}(X) \to Conf_{\tau,nsst}(B)/G.$$

is also a homeomorphism. □

REMARK 5.5. For the purposes of this paper we only need to know that $\mathcal{M}_{\tau,sst}(B) \neq \emptyset$ iff $\mathcal{M}_{n,\tau}(X) \neq \emptyset$.

Below are few more details concerning the symplectic nature of the moduli space $\mathcal{M}_{\tau,sst}(B)$ in the case when G is a complex Lie group. Recall that $\mathfrak{g} = \mathfrak{p} \oplus \mathfrak{k}$ and $\mathfrak{p} = \sqrt{-1}\mathfrak{k}$. For $\tau \in \Delta^n \subset \mathfrak{a}^n \subset \mathfrak{p}^n$ we identify each τ_i with the element $\alpha_i := \sqrt{-1}\tau_i$ of the Lie algebra $\mathfrak{k}$. We consider the product $M := \prod_{i=1}^n \mathcal{O}_i$ of the orbits $\mathcal{O}_i := Ad(K)(\alpha_i)$, $i = 1, ..., n$. The manifold M carries a natural symplectic structure which is invariant under the diagonal adjoint action of the group K. Let $f : M \to \mathfrak{k}$ be the *momentum mapping* of this action. Since the map f is given by the formula

$$f : (\beta_1, .., \beta_n) \mapsto \sum_{i=1}^n \beta_i,$$

we obtain an identification of the moduli space $\mathcal{M}_{n,\tau}(\mathfrak{p})$ of polygons in $\mathfrak{p}$ with the *symplectic quotient*:

$$\mathcal{M}_{\tau,sst}(B) = Conf_\tau(B)//G := M//G := \{\xi : f(\xi) = 0\}/K.$$

In the case when all τ_i belong to the cocharacter lattice L, this quotient is the same as the *Mumford quotient* of the projective variety $Conf_\tau$ by the group G.

5.2. The polyhedron $D_n(X)$

One of the main results of [**KLM1**] is a description of $D_n(X)$ (where X is an infinitesimal symmetric space) in terms of the Schubert calculus in the Grassmannians associated to complex and real Lie groups G (i.e. the quotients G/P where P is a maximal parabolic subgroup of G). Below we describe $D_n(X)$ for the three classes of metric spaces considered in the present paper. We first do it in the context of symmetric spaces of *noncompact type* (i.e. their deRham decomposition contains no Euclidean factor[1]) since the description in this class is more natural.

Let X be a symmetric space of noncompact type and G the identity component of its isometry group. The ideal boundary $\partial_{Tits}X$ is a spherical building modeled on a spherical Coxeter complex (S, W) with model spherical Weyl chamber $\Delta_{sph} \subset S$. We identify S with an apartment in $\partial_{Tits}X$. Let Δ denote the Euclidean Weyl chamber of X. We identify Δ_{sph} with $\partial_{Tits}\Delta$.

Let B be the stabilizer of Δ_{sph} in G. For each vertex ζ of $\partial_{Tits}X$ one defines the generalized Grassmannian $Grass_\zeta = G\zeta = G/P$. (Here P is the maximal parabolic subgroup of G stabilizing ζ.) It is a compact homogeneous space stratified into B-orbits called *Schubert cells.* Every Schubert cell is of the form $C_\eta = B\eta$ for a unique vertex $\eta \in W\zeta \subset S^{(0)}$ of the spherical Coxeter complex. The closures $\overline{C_\eta}$ are called *Schubert cycles.* They are unions of Schubert cells and represent well defined elements in the homology $H_*(Grass_\zeta, \mathbb{Z}_2)$.

For each vertex ζ of Δ_{sph} and each n-tuple $\vec{\eta} = (\eta_1, \dots, \eta_n)$ of vertices in $W\zeta$ consider the following homogeneous linear inequality for $\xi \in \Delta^n$:

$$(*_{\zeta;\vec{\eta}}) \qquad \sum_i \xi_i \cdot \eta_i \le 0$$

Here we identify the η_i's with unit vectors in Δ.

Let $I_{\mathbb{Z}_2}(G)$ be the subset consisting of all data $(\zeta, \vec{\eta})$ such that the intersection of the Schubert classes $[\overline{C}_{\eta_1}], \dots, [\overline{C}_{\eta_n}]$ in $H_*(Grass_\zeta, \mathbb{Z}_2)$ equals $[pt]$.

THEOREM 5.5 ([**KLM1**]). *$\Delta^n_{ss}(\partial_{Tits}X) \subset \Delta^n$ consists of all solutions ξ to the system of inequalities $(*_{\zeta;\vec{\eta}})$ where $(\zeta, \vec{\eta})$ runs through $I_{\mathbb{Z}_2}(G)$.*

REMARK 5.6. This system of inequalities depends on the Schubert calculus for the generalized Grassmannians G/P associated to the group G. It is one of the results of [**KLM2**] that the set of solutions depends only on the spherical Coxeter complex.

Typically, the system of inequalities in Theorem 5.5 is redundant. If G is a *complex* Lie group one can use the complex structure to obtain a smaller system of inequalities. In this case, the homogeneous spaces $Grass_\zeta$ are complex manifolds and the Schubert cycles are complex subvarieties and hence represent classes in *integral* homology. Let $I_{\mathbb{Z}}(G) \subset I_{\mathbb{Z}_2}(G)$ be the subset consisting of all data $(\zeta, \vec{\eta})$ such that the intersection of the Schubert classes $[\overline{C}_{\eta_1}], \dots, [\overline{C}_{\eta_n}]$ in $H_*(Grass_\zeta, \mathbb{Z})$ equals $[pt]$.

The following analogue of Theorem 5.5 was proven independently and by completely different methods in [**BeSj**] and in [**KLM1**]:

---

[1] In the context of Lie groups it corresponds to the case of semisimple algebraic groups.

THEOREM 5.6 (Stability inequalities). *$\Delta^n_{ss}(\partial_{Tits}X)$ consists of all solutions ξ to the system of inequalities $(*_{\zeta;\vec{\eta}})$ where $(\zeta, \vec{\eta})$ runs through $I_{\mathbb{Z}}(G)$.*

We now consider the general case when X is a symmetric space or a Euclidean building which splits as $X_0 \times X_1$, where X_0 is the flat deRham factor of X. In the case when $X = \mathfrak{p}$ is an infinitesimal symmetric space we consider the decomposition $X = X_0 \times X_1$ corresponding to the orthogonal decomposition $\mathfrak{p} = \mathfrak{p}_0 \oplus \mathfrak{p}_1$, where $\mathfrak{p}_0$ is the Lie algebra of the split part of the of the central torus of G (where $\mathfrak{g}$ is the Lie algebra of G). In this case we again refer to $\mathfrak{p}_0 := X_0$ as the Euclidean deRham factor of X. If $\mathfrak{p} = T_p(X')$, where X' is a nonpositively curved symmetric space, then the above decomposition of $\mathfrak{p}$ is the infinitesimal version of the splitting off the deRham factor of X'.

Let $E = E_0 \times E_1$ be the corresponding decomposition of the Euclidean Coxeter complex, the Weyl chamber $\Delta \subset E$ splits as $\Delta_0 \times \Delta_1$, where $\Delta_1 \subset E_1$ is the Weyl chamber for the action $W_{sph} \curvearrowright E_1$ and $\Delta_0 = E_0$. Let $D_n(X_i) \subset (\Delta_i)^n$ denote the side-lengths polyhedron for the space X_1. It is clear that

$$D_n(X_0) = \{(\sigma_1, ..., \sigma_n) : \sum_{i=1}^{n} \sigma_i = 0\}.$$

Then we get:

$$D_n(X) = D_n(X_0) \times D_n(X_1). \tag{5.2}$$

We refer the reader to Proposition 7.14 for the explanation.

Combining our results one obtains the following recipe for determining the polytope $D_n(X)$ for any of the spaces X as in Theorem 5.3 (i.e., infinitesimal symmetric spaces, symmetric spaces and Euclidean buildings): Given X, first of all, split X as $X_0 \times X_1$, where X_0 is the Euclidean deRham factor. In view of the formula (5.2) it suffices to describe $D_n(X_1)$; so we let $X := X_1$. Then find a complex semisimple Lie group G of noncompact type, whose spherical Coxeter complex is isomorphic to the one of X. Let X' be the symmetric space G/K associated with G. Using Schubert calculus for the Grassmannians G/P associated to G as in Theorem 5.6 above, compute the system of stability inequalities describing $\Delta^n_{ss}(\partial_{Tits}X') = D_n(X')$. The polytopes $D_n(X)$ and $D_n(X')$ are equal. We note that although the polyhedron $D_n(X)$ depends only on the spherical Coxeter group, the system of stability inequalities describing $D_n(X)$ depends on the root system. In fact in [**KLM1**] it is shown that the systems obtained for the root systems B_3 and C_3 are different (even though they have the same number of inequalities).

EXAMPLE 5.7. Suppose that G is a reductive algebraic over $\mathbb{Q}$ and $G^\vee$ is its Langlands' dual. Let $\mathfrak{g}$ and $\mathfrak{g}^\vee$ be the Lie algebras of G and $G^\vee$ and let $X = G/K, X^\vee$ be the corresponding symmetric spaces. Let $\mathfrak{g} = \mathfrak{k} \oplus \mathfrak{p}$ be the Cartan decomposition. Take a Cartan subalgebra $\mathfrak{a} \subset \mathfrak{p}$. The pairing $\langle\ ,\ \rangle$ induces an isomorphism $\mathfrak{a}^* \cong \mathfrak{a}^\vee$, where $\mathfrak{a}^\vee$ is a Cartan subalgera of $\mathfrak{g}^\vee$. We then identify $\mathfrak{a}$ with $\mathfrak{a}^*$ using the invariant metric. This gives us an isometry $f : \mathfrak{a} \to \mathfrak{a}^\vee$ which conjugates the spherical Coxeter group W_{sph} of $\mathfrak{g}$ to the spherical Coxeter group $W^\vee_{sph}$ of $\mathfrak{g}^\vee$. This isometry also carries a Euclidean Weyl chamber Δ of $W_{sph} \curvearrowright \mathfrak{a}$ onto a Euclidean Weyl chamber for the action $W^\vee_{sph} \curvearrowright \mathfrak{a}^\vee$. Thus f induces a bijection

$$f^n : D_n(X) \to D_n(X^\vee).$$

For instance, if G and $G^\vee$ are simple complex Lie groups of type B_ℓ and C_ℓ respectively, then the above construction provides an isometry $D_n(X) \cong D_n(X^\vee)$.

In the next subsection we will write down the inequalities 5.6 for the root system B_2 in the case $n = 3$.

5.3. The polyhedron for the root system B_2

In [**KLM1**] and [**KuLM**] the stability inequalities for the polyhedra $D_3(X)$ were computed for all X of rank 2 or 3 (the polyhedron for G_2 was computed in [**BeSj**]). We now give the example of the polyhedron for the root system B_2. This example will be useful to us later.

Since all symmetric spaces with the same root system give rise to the same polyhedron we may take $X = SO(5, \mathbb{C})/SO(5)$. The Weyl chamber Δ is given by

$$\Delta = \{(x, y) : x > y > 0\}.$$

Thus the Δ-lengths σ_i are vectors (x, y) in Δ. Since the root system has rank 2, Δ_{sph} has exactly two vertices ζ_1, ζ_2. Thus we get two generalized Grassmannians, $Grass_{\zeta_i}, i = 1, 2$. These Grassmannians are the spaces of isotropic lines and isotropic planes in $\mathbb{C}^4$. Thus the set of stability inequalities breaks into two subsystems, one for each ζ_i. This gives a system of 19 inequalities in addition to the inequalities defining the chamber Δ. Below we have dropped one of these 19 inequalities, which was implied by the inequalities defining the chamber.

The first subsystem of stability inequalities (corresponding to the Grassmannian of isotropic lines) is given by

$$x_i \leq x_j + x_k, \quad \{i, j, k\} = \{1, 2, 3\}$$
$$y_i \leq y_j + x_k, \quad \{i, j, k\} = \{1, 2, 3\}.$$

In order to describe the second system we let S be the sum of all the coordinates of the side-lengths, so

$$S = x_1 + y_1 + x_2 + y_2 + x_3 + y_3.$$

The second subsystem (corresponding to the Grassmannian of isotropic planes) is then given by

$$x_i + y_j \leq S/2, \quad 1 \leq i, j \leq 3.$$

To the above system of 18 inequalities we also have to add the "chamber inequalities":

$$x_i \geq y_i, i = 1, 2, 3 \text{ and } y_i \geq 0, i = 1, 2, 3.$$

In total we get 24 inequalities. Thus we have:

COROLLARY 5.8. *Suppose that X is a symmetric space with the (finite) Weyl group of type B_2. Then there exists a triangle in X whose Δ-side lengths are vectors $\sigma_i = (x_i, y_i) \in \Delta$, $i = 1, 2, 3$ if and only if σ_i's satisfy the above system of 18 inequalities.*

REMARK 5.9. Using a computer we have verified that the above system is minimal, the polyhedron D_3 is a cone over a compact polytope with 15 vertices and 24 top-dimensional faces.

Note that the Weyl chamber $\Delta \subset \mathfrak{a}$ determines a partial order on $\mathfrak{a}$:

$$\alpha \leq_\Delta \beta \iff \beta - \alpha \in \Delta.$$

One may ask if the "naive" triangle inequalities

$$\alpha \leq_\Delta \beta + \gamma$$

are satisfied by the Δ-side lengths of triangles in a symmetric space X with the Weyl chamber Δ. Below is a counter-example:

Consider the root system $B_2 = C_2$ and the vectors

$$\alpha = \begin{pmatrix} x_1 \\ y_1 \end{pmatrix}, \quad \beta = \begin{pmatrix} x_2 \\ y_2 \end{pmatrix} = \gamma = \begin{pmatrix} x_3 \\ y_3 \end{pmatrix} = \begin{pmatrix} t \\ t \end{pmatrix},$$

where $0 < t < y_1 < x_1 < 2t$. The reader will verify that the stability inequalities above are satisfied by the vectors α, β, γ, however the inequality

$$\alpha \leq_\Delta \beta + \gamma$$

fails since it would imply that $0 < x_1 - y_1 < (x_2 - y_2) + (x_3 - y_3) = 0$.

Nevertheless, one can rewrite the system of stability inequalities for the root system $B_2 = C_2$ in terms of the *root cone*

$$\Delta^* = \{(x, y) : x \geq 0, x + y \geq 0\} = \{v \in \mathbb{R}^2 : \forall u \in \Delta, u \cdot v \geq 0\}$$

as follows:

$$(\alpha_1, \alpha_2, \alpha_3) \in D_3(X) \iff$$
$$\alpha_1, \alpha_2, \alpha_3 \in \Delta,$$

and

$$\alpha_i \leq_{\Delta^*} \alpha_j + \alpha_k,$$
$$\tau(\alpha_i) \leq_{\Delta^*} \tau(\alpha_j) + \alpha_k$$

for all i, j, k such that $\{i, j, k\} = \{1, 2, 3\}$. Here $\tau(x, y) = (y, x) \in W_{sph}$.

CHAPTER 6

Polygons in Euclidean Buildings and the Generalized Invariant Factor Problem

Let X be a thick Euclidean building modeled on a discrete Coxeter complex (E, W_{aff}). As in section 4.4, let L be a lattice in E which contains the translation subgroup L_{trans} of W_{aff} and which normalizes W_{aff}. Note that L acts by automorphisms on the Coxeter complex (E, W_{aff}). For the algebraic applications we would like to determine for which $\vec{\tau} \in D_n(X) \cap L^n$, there exists a polygon in X with L-integral side-lengths $\vec{\tau}$ and the first vertex at a distinguished special vertex $o \in X$. In other words, we are interested in the image of the map

$$\iota : D_n^{ref,L}(X) \to D_n(X) \cap L^n. \tag{6.1}$$

In this chapter we will show that in general the map (6.1) is not onto (section 6.2): The counterexamples are based on the idea of *folding* triangles into apartments (see section 6.1). In the subsequent chapter 7 we will prove "positive results": Some of them guarantee that the map (6.1) is onto for certain pairs (W_{aff}, L), the other results establish sufficient conditions for elements of $D_n(X) \cap L^n$ to belong to the image of ι.

6.1. Folding polygons into apartments

In this section we describe a construction which produces billiard triangles in an apartment from triangles in a building .

Suppose that $\Delta(x, y, z)$ is a triangle in a (Euclidean or spherical) building B. In general it is not contained in an apartment. However (see [**KLM2**, §3.2]) there exists a finite subdivision of the edge $\overline{xy}$ by points $x_0 = x, x_1, ..., x_{k-1}, x_k = y$ such that each geodesic triangle $\Delta(z, x_i, x_{i+1})$ is contained in an apartment in B.

REMARK 6.1. It is easy to see that there is a uniform upper bound on the number k which depends only on the Coxeter group W.

For each i, let $a_i \subset B$ be an apartment containing $\Delta(z, x_i, x_{i+1})$. We will identify a_0 with the model apartment A. We will produce points x_i' in the first apartment a_0 such that the triangles $\Delta(z, x_i', x_{i+1}')$ are congruent to the triangles $\Delta(z, x_i, x_{i+1})$ via apartment isomorphisms $\alpha_i : a_i \to A$. This is done inductively as follows. We start with $x_0' = x_0$ and $x_1' = x_1$. Suppose that x_i' has been constructed. To find x_{i+1}', choose $\alpha_i : a_i \to a$ so that it carries $\overline{zx_i}$ to $\overline{zx_i'}$. Put $x_{i+1}' = \alpha_i(x_{i+1})$. The procedure yields a *billiard triangle* in the apartment A consisting of two geodesic sides $\overline{zx_0'}$ and $\overline{zx_k'}$ with the same refined lengths as the corresponding sides of the original triangle $\Delta(x, y, z)$, and one piecewise geodesic path $x_0'x_1' \dots x_k'$. The points $x_1', ..., x_{k-1}'$ are the *break points* of the broken side of this billiard triangle.

REMARK 6.2. If z is a special vertex of X, then, by projecting the apartment A to a Weyl chamber Δ (with the tip at z), we can assume that the folded triangle is contained in Δ.

Consider the spherical building $Y_i := \Sigma_{x_i}(B)$ whose Coxeter complex is

$$(\Sigma_{x_i'} A, W_i),$$

where the group W_i is the stabilizer of x_i in W. Let θ_i be the canonical projection of Y_i to the Weyl chamber in this building.

REMARK 6.3. Notice that W_i may be smaller than W_{sph}.

Since the refined lengths of $\overline{x_i x_{i+1}}$ and $\overline{x_i' x_{i+1}'}$ are equal, it follows that

$$\theta_i(\overrightarrow{x_i x_{i-1}}) = \theta_i(\overrightarrow{x_i' x_{i-1}'})$$

and

$$\theta_i(\overrightarrow{x_i x_{i+1}}) = \theta_i(\overrightarrow{x_i' x_{i+1}'}).$$

Because the directions $\overrightarrow{x_i x_{i-1}}$ and $\overrightarrow{x_i x_{i+1}}$ in the spherical building $Y_i = \Sigma_{x_i}(B)$ are antipodal, the properties of the canonical projection θ_i (see section 4.2) imply that the directions $\overrightarrow{x_i' x_{i-1}'}$ and $\overrightarrow{x_i' x_{i+1}'}$ in the spherical Coxeter complex $(\Sigma_{x_i'} A, W_i)$ are antipodal modulo the action of W_i.

DEFINITION 6.4. A broken triangle $T \subset A$ with two geodesic sides $\overline{z x_0'}$ and $\overline{z x_k'}$ and one piecewise geodesic path $x_0' x_1' \dots x_k'$ is a *billiard triangle* if at every break point x_i', the directions $\overrightarrow{x_i' x_{i-1}'}$ and $\overrightarrow{x_i' x_{i+1}'}$ in the spherical Coxeter complex $\Sigma_{x_i'} A$ are antipodal modulo the action of the stabilizer W_i of x_i' in the Coxeter group of A.

We note that each broken side $x_0' x_1' \dots x_k'$ of a billiard triangle can be *straightened* in the model apartment A, i.e. there exists a geodesic segment $\overline{x_0'' x_k''} \subset A$ (a *straightening* of $x_0' x_1' \dots x_k'$) such that $x_0'' = x_0'$, the metric length of $\overline{x_0'' x_k''}$ is the same as of $x_0' x_1' \dots x_k'$, and the direction of $\overline{x_0'' x_k''}$ at x_0' is the same as the direction of $\overline{x_0' x_1'}$.

A billiard triangle T in the model apartment can be unfolded to a geodesic triangle in the building if and only if, for each break point x_i', $0 < i < k$, the following holds. Let $\xi_i', \eta_i', \zeta_i' \in \Sigma_{x_i'} A$ be the directions towards x_{i-1}', x_{i+1}', z. The necessary and sufficient condition is:

CONDITION 6.5. For each i there exists a triangle $\Delta(\xi_i, \eta_i, \zeta_i)$ in the spherical building $Y_i = \Sigma_{x_i'} B$ so that $\angle(\xi_i, \eta_i) = \pi$ and the refined lengths of $\overline{\xi_i \zeta_i}$ and $\overline{\eta_i \zeta_i}$ are the same as for $\overline{\xi_i' \zeta_i'}$ and $\overline{\eta_i' \zeta_i'}$ respectively.

REMARK 6.6. A necessary condition for existence of a triangle $\Delta(\xi_i, \eta_i, \zeta_i)$ is that $\angle(\xi_i', \zeta_i') + \angle(\eta_i', \zeta_i') \geq \pi$, which is just the usual (metric) triangle inequality in the spherical building Y_i.

LEMMA 6.7. *Suppose that we have an apartment a' in a Euclidean building X and a billiard triangle $T' \subset a'$, T' has geodesic sides $\overline{z' x_0'}, \overline{z' x_k'}$ and the broken side $x_0' x_1' \dots x_k'$. Suppose that for each break point $x_i', i = 1, ..., k-1$, there is a wall $H_i \subset a'$ through x_i' which weakly separates*[1] *$\{x_{i-1}', x_{i+1}'\}$ from the vertex z'. Assume*

---

[1] I.e., H_i separates a' into half-apartments $a_+', a-'$ such that z is in the closure of a_+' and x_{i-1}', x_{i+1}' are in the closure of a_-'.

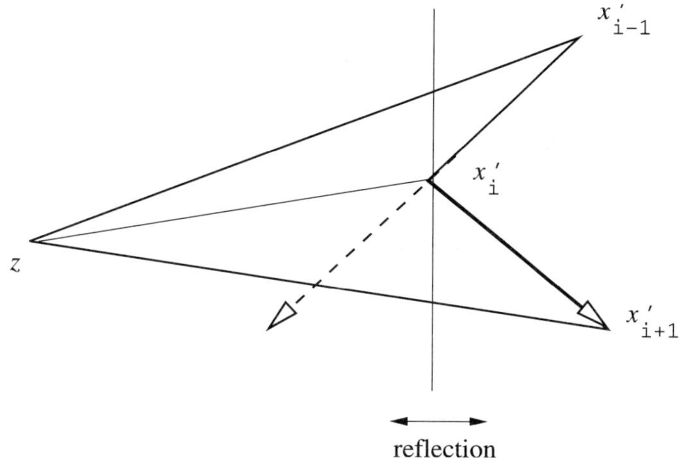

FIGURE 1. *A billiard triangle.*

that the reflection w in the wall H_i carries the direction $\overrightarrow{x'_i x'_{i+1}}$ to the direction antipodal to $\overrightarrow{x'_i x'_{i-1}}$. Then there exists a geodesic triangle $T = \Delta(x, y, z) \subset X$ so that

$$\sigma_{ref}(\overline{z'x'_0}) = \sigma_{ref}(\overline{zx}), \sigma_{ref}(\overline{z'x'_k}) = \sigma_{ref}(\overline{zy})$$

and the refined side-length of $\overline{xy}$ is the same as for the straightening of the broken side $x'_0 x'_1 \dots x'_k$.

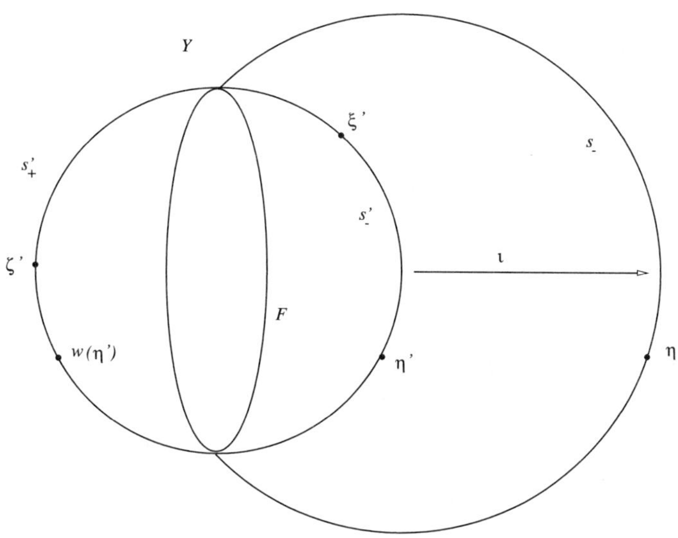

FIGURE 2. *Unfolding.*

Proof: We prove the lemma by verifying Condition 6.5 for each $i = 1, \dots, k-1$. The spherical building $Y := \Sigma_{x'_i} X$ contains the apartment $s' = \Sigma_{x'_i}(a')$ and the directions ξ', ζ', η' of the geodesic segments

$$\overline{x_i' x_{i-1}'}, \overline{x_i' x_{i+1}'}, \overline{x_i' z'}.$$

The wall $F := \Sigma_{x_i'}(H_i)$ in the spherical apartment s' separates s' into half-apartments s'_+, s'_-, so that $cl(s'_+)$ contains ζ' and $cl(s'_-)$ contains the directions ξ', η'. Because the building Y is thick, there exists an apartment $s \subset Y$ which intersects s' along the half-apartment s'_+. Let $s_- := cl(s \setminus s'_+)$. There exists an isomorphism of Coxeter complexes $\iota : s' \to s$ which restricts to the identity on s_+. Set $\eta := \iota(\eta'), \zeta := \zeta', \xi := \xi'$. Clearly,

$$\sigma_{ref}(\overline{\zeta\xi}) = \sigma_{ref}(\overline{\zeta'\xi'}), \quad \sigma_{ref}(\overline{\zeta\eta}) = \sigma_{ref}(\overline{\zeta'\eta'}).$$

It remains to verify that $\angle(\xi, \eta) = \pi$, i.e. that the points ξ and η are antipodal. Note that we also have the third apartment $s'' = s'_- \cup s_-$ and the isomorphism of Coxeter complexes

$$j : s' \to s''$$

which is the identity on s'_-, see Figure 2. The restriction of j to s'_+ equals $\iota \circ w$. Thus j carries $w(\eta')$ (which is antipodal to ξ' by assumption) to the point η, and $j(\xi') = \xi$. Thus the points η, ξ are antipodal. □

We now state a conjecture describing unfoldable billiard triangles. Let $\Delta \subset E$ be a Weyl chamber with the tip at a special vertex o. We first define a *weak LS path* between two special vertices $x = x_0, y = x_k \in \Delta$ to be a broken geodesic $x_0 x_1 \dots x_k \subset \Delta$ which satisfies Littelmann's axioms for an LS path in [**L2**, §4] except we do not require $dist(\lambda_{i-1}, \lambda_i) = 1$ in the definition of an a-chain. A billiard triangle is a *Littelmann triangle* if the broken side $x_0 x_1 \dots x_k$ is an LS path. We say that a billiard triangle in Δ with the geodesic sides $\overline{ox_0}, \overline{ox_k}$ and a broken side $x_0 x_1 \dots x_k$ is a *generalized Littelmann triangle* if $x_0 x_1 \dots x_k$ is a weak LS path. We recall (see [**L2**]) that there exists a Littelmann triangle with geodesic sides $\overline{ox_0}, \overline{ox_k}$ and the broken side $x_0 x_1 \dots x_{k-1} x_k$ iff $V_\gamma \subset V_\alpha \otimes V_\beta$ where $\alpha := \sigma(\overline{ox_0}), \gamma = \sigma(\overline{ox_k})$ and β is the Δ-length of the straightening of the broken geodesic $x_0 x_1 \dots x_{k-1} x_k$, provided that α, β, γ are characters of the split torus $\underline{T}^\vee \subset \underline{G}^\vee(\mathbb{C})$. Here $V_\alpha, V_\beta, V_\gamma$ are irreducible representations of the group $\underline{G}^\vee(\mathbb{C})$, cf. §1.

CONJECTURE 6.8. A billiard triangle in Δ with the sides $\overline{ox_0}, \overline{ox_k}$ and

$$x_0 x_1 \dots x_{k-1} x_k$$

and with the special vertices $0, x_0, x_k$, is unfoldable iff it is a generalized Littelmann triangle.

6.2. A Solution of Problem Q2 is not necessarily a solution of Problem Q3

In this section we first construct an example of a discrete thick Euclidean building X (modeled on discrete Euclidean Coxeter complex with the root system R of type B_2) and a triangle $P \subset X$ with the Δ-side lengths $\tau_1, \tau_2, \tau_3 \in L = Q(R^\vee)$ so that:

1. The vertices of P are at vertices of X.
2. There is no triangle $P' \subset X$ with vertices at special vertices of X and the Δ-lengths τ_1, τ_2, τ_3.

In terms of our basic algebra problems, (α, β, γ) is a solution of Problem **Q2** but not of Problem **Q3** for the simply-connected group $\underline{G} = Spin(5)$,

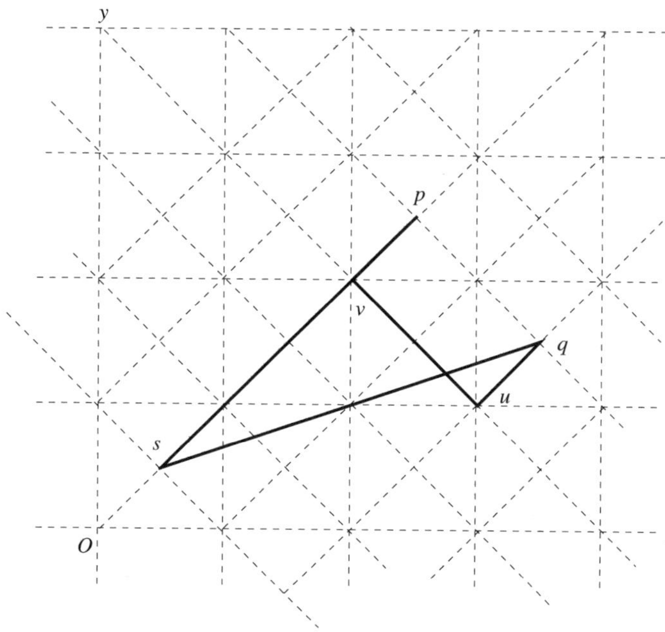

FIGURE 3. *A folded triangle.*

We next describe (without proof) analogous counterexamples for the root systems $C_\ell, \ell \geq 2$ and $L = Q(R^\vee)$. In terms of our basic algebra problems, this shows that there are solutions of Problem **Q2** that are not solutions to Problem **Q3** for the simply-connected groups $\underline{G} = Sp(2\ell), \ell \geq 2$.

EXAMPLE 6.9. Let X be a thick Euclidean building with associated discrete Euclidean Coxeter complex (E, W_{aff}) of type B_2. Then the map (6.1) is not surjective for $n = 3$ and $L = Q(R^\vee)$.

We recall that the simple roots for the root system B_2 are $(1,-1)$ and $(0,1)$ and the simple coroots are $(1,-1)$ and $(0,2)$. The folded triangle T is represented in Figure 3. It has three vertices: s, p and q and two geodesic sides: $\overline{sp}$ and $\overline{sq}$. The third side is the broken geodesic segment which consists of three pieces: $\overline{pv}, \overline{vu}$ and $\overline{uq}$. The vectors in the Weyl chamber $\Delta = \{(x,y) \in \mathbb{R}^2 : 0 \leq y \leq x\}$ which represent the corresponding Δ-side lengths are $(2,2)$, $(3,1)$ and $(2,2)$, where $(2,2)$ represents the broken side. To see that T can be unfolded into a geodesic triangle in the corresponding building X one can either use Lemma 6.7 or simply verify that the vectors $(2,2)$, $(3,1)$ and $(2,2)$ satisfy the inequalities of Section 5.3. Note that the vectors $(3,1)$ and $(2,2)$ belong to the coroot lattice.

We now prove that there is no triangle $\Delta(s'', p'', q'')$ in the building X whose Δ-side lengths are the vectors $(2,2)$, $(3,1)$ and $(2,2)$ and whose vertices are special vertices of X. Suppose that such triangle exists. As described in section 6.1, we subdivide the side $\overline{p''q''}$ and fold $\Delta(s'', p'', q'')$ to a billiard triangle T' in the model apartment A. Without loss of generality we can assume that the side $\overline{s''p''}$ folds onto the geodesic segment $\overline{oc}$, where o is the origin in A and $c = (2,2)$. By Remark 6.2 we can assume that this triangle is contained in the Weyl chamber $\Delta = \{(x,y) : x \geq y \geq 0\}$. Hence the side $\overline{s''q''}$ folds onto the geodesic segment $\overline{oa}$

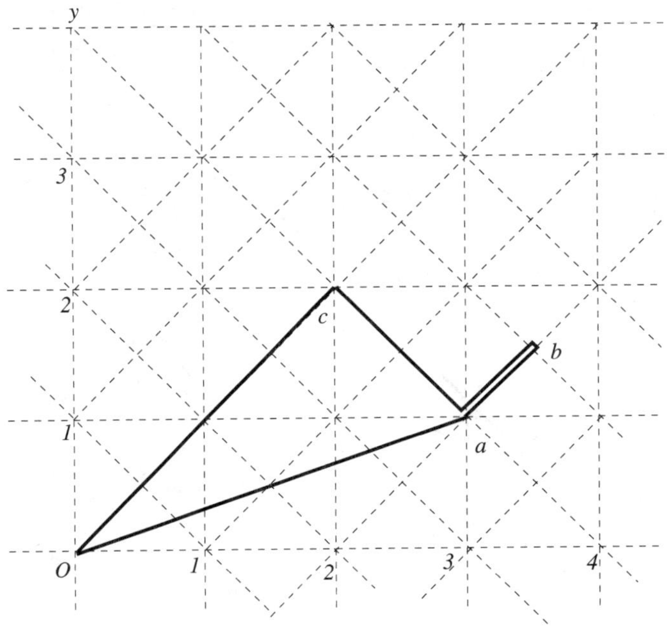

FIGURE 4. *A billiard triangle.*

where $a = (3,1)$. Note that $d(a,c)$ is strictly less than $2\sqrt{2}$, which is the magnitude of the vector $(2,2)$. Thus the broken geodesic segment $f(\overline{p''q''})$ (which is the image of $\overline{p''q''}$ under folding) has to have at least one break point. Next, observe that the only break points in $f(\overline{p''q''})$ can occur at the vertices of the affine Coxeter complex. Any vertex other than one in the set

$$S := \{(2,2), (3,1), (2.5, 0.5), (1.5, 1.5),$$
$$(2.5, 1.5), (2.5, 2.5), (3.5, 0.5), (3.5, 1.5)\}$$

would be too far from a, c for a billiard triangle to exist. We note that all points in $S \setminus \{a, c\}$ are nonspecial vertices of the Coxeter complex. Thus the only break in $f(\overline{p''q''})$ which can occur at such a point is "backtracking".

Below we exclude breaks at various points of S and will leave the rest of the possibilities to the reader, since the arguments are similar. Suppose that the broken geodesic path $f(\overline{p''q''})$ is the concatenation of the geodesic segments

$$\overline{ca}, \overline{ab}, \overline{ba},$$

where $b = (3.5, 1.5)$, see Figure 4. The point b is one of two break points of $f(\overline{p''q''})$. At the link $\Sigma_b(A)$ consider the directions:

$$\xi', \eta', \zeta'$$

towards the points a, a and o. Then

$$\angle(\zeta', \xi') + \angle(\zeta', \eta') = 2\angle(\zeta', \xi') < \pi.$$

Hence, according to Remark 6.6, the billiard triangle T' cannot be unfolded into a geodesic triangle $\Delta(s'', p'', q'') \subset X$. Contradiction.

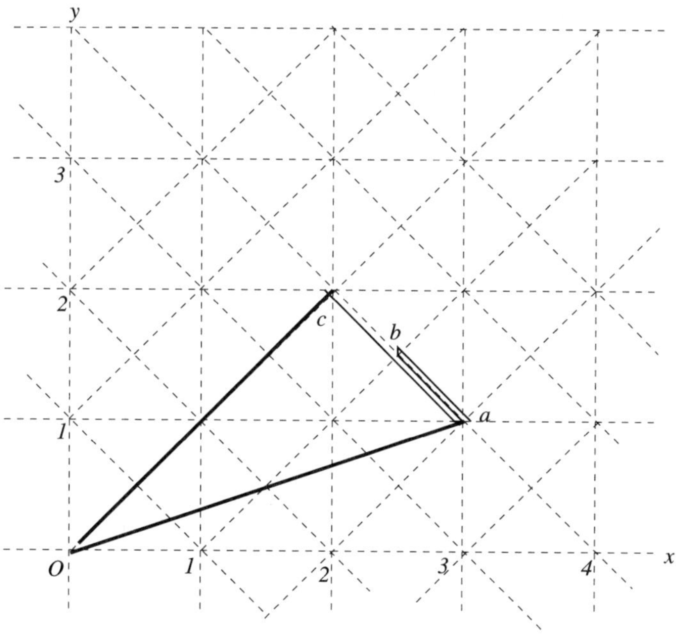

FIGURE 5. *A billiard triangle.*

Lastly, consider a break point at the vertex $b = (2.5, 2.5)$; the broken geodesic path $f(\overline{p''q''})$ is the concatenation of the geodesic segments

$$\overline{ca}, \overline{ab}, \overline{ba},$$

see Figure 5. At the link $\Sigma_a(A)$ of the break point a consider the directions:

$$\xi', \eta', \zeta'$$

towards the points c, b and o. Again,

$$\angle(\zeta', \xi') + \angle(\zeta', \eta') = 2\angle(\zeta', \xi') < \pi,$$

contradiction. (Note that in this example we do not get a contradiction by considering the link of the break point b.) □

REMARK 6.10. Consider the usual embedding $R_2 \to R_\ell$ of the root system $R_2 = B_2$ to $R_\ell = B_\ell$. The Δ-side length vectors $(2,2)$, $(3,1)$ and $(2,2)$ correspond to the vectors

$$\alpha = \beta = (2, 2, 0, ..., 0), \gamma = (3, 1, 0, ..., 0) \in L = Q(R_\ell^\vee).$$

One can show that for each $\ell \geq 3$ and the appropriate discrete building X, the vectors α, β, γ **belong to the image** of $\iota : D_3^{ref,L}(X) \to D_3(X)$.

Consider now a thick building X'_ℓ modeled on the discrete Euclidean Coxeter complex of type $R'_\ell := C_\ell$, $\ell \geq 3$. Set $L = Q((R'_\ell)^\vee)$. To obtain examples of triples of vectors which do not belong to the image of

$$D_3^{ref,L}(X'_\ell) \overset{\iota}{\hookrightarrow} D_3(X'_\ell)$$

we do the following:

Choose the vectors $\alpha' = \beta' = (2,0)$ and $\gamma' = (2,1)$. Then α', β', γ' are in the coroot lattice L of C_2. We claim this choice of side-lengths gives a solution of Problem **Q2** that is not a solution of Problem **Q3** for the root system C_2.

LEMMA 6.11. *There exists an isomorphism of algebraic groups*

$$\phi : Spin(5) \longrightarrow Sp(4)$$

carrying a split torus of $Spin(5)$ *to a split torus of* $Sp(4)$ *such that the induced map on Cartan subalgebras (relative to the coordinates of* [**Bo**, pg. 252-255]*), is given by the matrix*

$$\frac{1}{2}\begin{bmatrix} 1 & 1 \\ 1 & -1 \end{bmatrix},$$

Proof: Let $\phi : Spin(5) \to SL(4)$ be the spin representation. By Theorem G (b) of [**Sam**], the spin representation of $Spin(2n+1)$ is symplectic if and only if either $n \equiv 0 \bmod 4$ or $n \equiv 1 \bmod 4$. Hence, the image of ϕ lies in $Sp(4)$. If follows easily that ϕ is an isomorphism. Let $\epsilon_i, i = 1, 2$ be the coordinate functionals as in [**Bo**], loc. cit. Since the weights of the spin representation relative to the basis of the dual of the Cartan subalgebra are $(1/2, 1/2), (1/2, -1/2), (-1/2, 1/2)$ and $(-1/2, -1/2)$ we find that the pull-back by the map, induced on the duals of the Cartan subalgebras, of the coordinate functional $\epsilon_1 = (1,0)$ is $(1/2, 1/2)$ and the pull-back of the coordinate functional $\epsilon_2 = (0,1)$ is $(1/2, -1.2)$. Thus the above matrix is the matrix of the map on the duals. Since it is symmetric is also the matrix of the map on the Cartan subalgebras. □

Since the vectors α, β, γ (from the previous example) map to the vectors $\alpha' = \beta'$ and $\gamma' = (2,1)$ respectively, the claim follows.

Now consider the natural embedding of root systems $C_2 \hookrightarrow C_\ell = R_\ell$, $\ell \geq 2$. One can verify (similarly to the arguments presented in Example 6.9) that the vectors $\alpha' = \beta', \gamma' \in L = Q(R_\ell^\vee)$ satisfy the property that

$$(\alpha', \beta', \gamma') \in D_3^L(X_\ell) \setminus \iota(D_3^{ref,L}(X_\ell)),$$

where X_ℓ is a discrete Euclidean building modeled on the Euclidean Coxeter complex associated with the root system C_ℓ. Thus the triple $(\alpha', \beta', \gamma')$ is a solution of Problem **Q2** for $Sp(2\ell, \mathbb{C})$ but not a solution of Problem **Q3** for $Sp(2, \mathbb{K})$ where $\mathbb{K}$ is an nonarchimedean local field with value group $\mathbb{Z}$.

CHAPTER 7

The Existence of Fixed Vertices in Buildings and Computation of the Saturation Factors for Reductive Groups

As we have seen in the previous chapter, the map

$$\iota : D_n^{ref,L}(X) \to D_n(X) \cap L^n. \tag{7.1}$$

in general is not surjective. The goal of this chapter is to find conditions on the root systems, etc., which would guarantee existence of polygons P in Euclidean buildings X with the prescribed L-integral Δ-side lengths and vertices at the vertices of X. Moreover, we will find conditions under which the vertices of P are in $G \cdot o$, where $o \in X$ is a certain special vertex. We will also see that the image of ι is always contained in the set

$$D_n^{L,0}(X) = \{(\tau_1, ..., \tau_n) \in D_n(X) \cap L^n : \sum_{i=1}^{n} \tau_i \in Q(R^\vee) = L_{trans}\}.$$

We will show that for each data (W_{aff}, L) (where $L_{trans} \subset L \subset N_{aff}$), there exists a natural number k such that for each $\tau \in D_n(X) \cap L^n$, the vector $k\tau$ belongs to the image of the map (6.1). We will compute the *saturation factors* k for various classes of (W_{aff}, L). In few cases we are fortunate and $k = 1$, i.e. the map (7.1) is onto. We then apply these results to the algebra Problem **Q3**.

7.1. The saturation factors associated to a root system

In this section we define and compute *saturation factors* associated with root systems.

DEFINITION 7.1. Let (E, W_{aff}) be a Euclidean Coxeter complex, $W_{aff} = W_{R,\mathbb{Z}}$. We define the *saturation factor* k_R for the root system R to be the least natural number k such that $k \cdot E^{(0)} \subset E^{(0),sp} = N_{aff} \cdot o$. The numbers k_R for the irreducible root systems are listed in the table (7.3).

Below we explain how to compute the saturation factors k_R. First of all, it is clear that if the root system R is reducible and $R_1, ..., R_s$ are its irreducible components, then $k_R = LCM(k_{R_1}, ..., k_{R_s})$, where LCM stands for the *least common multiple.* Henceforth we can assume that the system R is reduced, irreducible and $n = \dim(V)$. Then the affine Coxeter group W_{aff} is discrete, acts cocompactly on E and its fundamental domain (a *Weyl alcove*) is a simplex.

Let $\{\alpha_1, ..., \alpha_n\}$ be the collection of simple roots in R (corresponding to the positive Weyl chamber Δ) and $\alpha_0 := \theta$ be the highest root. Then

$$\theta = \sum_{i=1}^{n} m_i \alpha_i. \tag{7.2}$$

We can choose as a Weyl alcove C for W_{aff} the simplex bounded by the hyperplanes $H_{\alpha_j,0}, H_{\theta,1}$, $j = 1, ..., n$. The vertices of C are: $o = x_0$ (the origin) and the points $x_1, ..., x_n$. Each $x_i, i \neq 0$, belongs to the intersection of the hyperplanes $H_{\alpha_0,1}, H_{\alpha_j,0}$, $1 \leq j \neq i \leq n$. The set of values (mod $\mathbb{Z}$) of the linear functionals α ($\alpha \in R$) on the vertex set $E^{(0)}$ of the Coxeter complex, equals $\{\alpha_i(x_i) : i = 1, ..., n\}$. Note that $1 = \theta(x_i) = m_i\alpha_i(x_i)$ where the numbers m_i are the ones which appear in the equation (7.2). Thus $\alpha_i(x_i) = \frac{1}{m_i}$.

LEMMA 7.2. $k_R = LCM(m_1, ..., m_n)$.

Proof: We have: $\alpha_i(kx_i) \in \mathbb{Z}$ for each i, which in turn implies that $\alpha(kx_i) \in \mathbb{Z}$ for all $\alpha \in R, i = 1, ..., n$. Hence $\alpha(kE^{(0)}) \subset \mathbb{Z}$ for each $\alpha \in R$. Since $N_{aff} = P(R^\vee)$, this proves that $kE^{(0)} \subset N_{aff} \cdot o$. If $k \in \mathbb{N}$ is such that $k \cdot E^{(0)} \subset P(R^\vee)$, then m_i divides k for each $i = 1, ..., n$. □

In our paper we will also need a generalization of the numbers k_R, which we discuss for the rest of this section. (The reader who is interested only in simply-connected groups can ignore this material.) We again consider a general reducible root system R. Suppose that L' is a subgroup of N_{aff} containing the lattice $L_{trans} = Q(R^\vee)$; we will assume that L' acts as a *lattice* on E (i.e. a discrete cocompact group). Set $L := L' \cap V_1 \oplus ... \oplus V_s$, where V_i are the vector spaces underlying E_i. We note that since $L_{trans} \subset L$ and L_{trans} acts as a lattice on $E_1 \oplus ... \oplus E_s$, the discrete group L also acts as a lattice on $E_1 \oplus ... \oplus E_s$.

Let p_i denote the orthogonal projections $E \to E_i$. Consider the images L_i of L under the projections p_i $(i = 0, ..., s)$; since $L \subset N_{aff}$ and $p_i(N_{aff}) = N^i_{aff}$, we have the inclusions

$$L^i_{trans} \subset L_i \subset N^i_{aff}, i = 1, ..., s,$$

where L^i_{trans} is the translation subgroup of W^i_{aff}.

EXAMPLE 7.3. Suppose that (E, W_{aff}) is the Coxeter complex associated with the root system of the group $GL(n)$. Then $E = \mathbb{R}^n$, $L' = L_{GL(n)} = \mathbb{Z}^n$ is the cocharacter group of the maximal torus $\underline{T}$ (represented by diagonal matrices) in $GL(n)$. The group L' is generated by the cocharacters $e_i = (0, ..., 0, 1, 0...0)$ (1 is on the i-th place). The coroot lattice $Q(R^\vee)$ is generated by the simple coroots $\alpha_i^\vee = e_i - e_{i+1}, i = 1, ..., n-1$. The metric on E is given by the trace of the product of matrices. We have the decomposition $E = E_0 \times E_1$ where E_0 is 1-dimensional and is spanned by the vector $e = e_1 + ... + e_n$, and the space E_1 is the kernel of the map

$$tr : (x_1, ..., x_n) \mapsto \sum_{i=1}^{n} x_i.$$

Thus E_1 is the (real) Cartan subalgebra of the Lie algebra $\mathfrak{sl}(n)$ of $SL(n)$, the derived subgroup of $GL(n)$.

The projection $p_1 : E \to E_1$ is given by $p_1(u) = u - \frac{1}{m}tr(u)e$. The group W_{aff} equals W^1_{aff}, which acts on E_1 as the Euclidean Coxeter group with the extended Dynkin diagram of type $\tilde{A}_{n-1}$.

The intersection $L = L' \cap E_1 = Q(R^\vee)$, where $R^\vee$ is a root system contained in E_1. It is the coroot system of the Lie algebra $\mathfrak{sl}(n)$ The projection $L_1 = p_1(L')$ is $P(R^\vee)$, the coweight lattice of the Lie algebra $\mathfrak{sl}(n)$.

Consider the group of isometries $\tilde{W}$ generated by elements of W_{aff} and L. Then $\tilde{W}$ is a Euclidean Coxeter group with the linear part W_{sph} and translation part L, $\tilde{W} = W_{sph} \ltimes L$. Since $\tilde{W}^i = W^i_{sph} \ltimes L_i$ normalizes W^i_{aff}, for each i we get the induced action of the finite abelian group $F_i := \tilde{W}^i/W^i_{aff} \cong L_i/L^i_{trans}$ on the Weyl alcove a_i of W^i_{aff}.

DEFINITION 7.4. A face $\delta_i \subset a_i$ of a Weyl alcove a_i of W^i_{aff}, will be called *L_i-admissible* if there exists an element $g \in F_i$ which preserves δ_i and $\langle g \rangle$ acts transitively on its vertices ($i = 1, ..., s$).

Note that in the case $L_i = L^i_{trans}$, the only L_i-admissible simplices are the vertices of δ_i.

DEFINITION 7.5. For each pair of groups (W^i_{aff}, L_i), define the *saturation factor* $k_i = k(W^i_{aff}, L_i) \in \mathbb{N}$, to be the smallest natural number k_i such that for each L_i-admissible face $c \subset a_i$, the multiple of its barycenter $k_i b_c$, belongs to $E_i^{(0),sp} = N^i_{aff} \cdot o$. We let $k(W_{aff}, L) = k(W_{aff}, L')$ denote $LCM(k_1, ..., k_s)$. In the case $L = N_{aff}$ we will use the notation k_w for $k(W_{aff}, L)$.

We note that in the case $L = L_{trans}$ we get $k(W_{aff}, L) = k_R$ and if L is the weight lattice, $L = P(R^\vee)$ we have $k(W_{aff}, L) = k_w$.

Our next goal is to compute the saturation factors for various irreducible root systems and various lattices L.

We again assume that the root system R is irreducible and that its rank n equals $\dim(V)$. Note that the group $F = L/L_{trans}$ acts by automorphisms on the extended Dynkin diagram $\tilde{\Gamma}$ of the root system R (since F acts on the Weyl alcove a which is uniquely determined by the labelled graph $\tilde{\Gamma}$, whose nodes correspond to the faces of a). For $i = 1, ..., n$ we mark the i-th node (corresponding to α_i) of $\tilde{\Gamma}$ with the natural number m_i which appears the formula for the highest root (7.2). We mark the 0-th node of $\tilde{\Gamma}$ (corresponding to θ) with 1. Then the automorphisms of $\tilde{\Gamma}$ preserve this labelling; the action of the full group N_{aff} on $\tilde{\Gamma}$ is transitive on the set of all the nodes labelled by 1. Not all automorphisms of $\tilde{\Gamma}$ can be induced by F even if one takes L as large as possible, i.e. $L = N_{aff}$. Recall that the action on a comes from the action of $\tilde{W}$ by conjugation on W_{aff}; this action induces inner automorphisms of the spherical Weyl group W_{sph}. Thus, if g is an automorphism of $\tilde{\Gamma}$ induced by an element of F and g fixes a vertex with the label 1, then g acts trivially on $\tilde{\Gamma}$. This does not completely determine the image of N_{aff} in $Aut(\tilde{\Gamma})$ but it will suffice for the computation of the saturation factors.

Here is the procedure for computing the saturation factor $k = k(W_{aff}, L)$. Given $g \in F$ (including the identity) consider the orbits of $\langle g \rangle$ in the vertex set of the graph $\tilde{\Gamma}$. Here and in what follows $\langle g \rangle$ denotes the cyclic subgroup of $\mathrm{Isom}(E)$ generated by g. Let $\overline{\mathcal{O}} = \{\overline{x}_{i_1}, ..., \overline{x}_{i_t}\}$ be such an orbit. This orbit corresponds to the orbit $\mathcal{O} = \{x_{i_1}, ..., x_{i_t}\}$ of $\langle g \rangle$ on the vertex set of the Weyl alcove. Take the barycenter

$$b(\mathcal{O}) = \frac{1}{t}\sum_{j=1}^{t} x_{i_j}$$

of the corresponding vertex set (also denoted $\mathcal{O}$) of the Euclidean Coxeter complex. For the point $b = b(\mathcal{O})$ compute the rational numbers $\alpha_i(b), i = 0, ..., n$.

Then find the GCD (the greatest common denominator) of the rational numbers $\alpha_i(b), i = 0, ..., n$, call it $k_{\mathcal{O}}$. Finally, let

$$k := LCM(\{k_{\mathcal{O}}, \text{ where } \mathcal{O} \text{ runs through all orbits of all } \langle g \rangle \subset F\}).$$

REMARK 7.6. Instead of taking all $g \in F$ it is enough to consider representatives of their conjugacy classes in $\tilde{W}/W_{aff}$ (under the conjugation by the full automorphism group of $\tilde{\Gamma}$).

It is clear that the number k computed this way satisfies the required property:

1. For the barycenter b of each L-admissible face of a, the multiple kb belongs to the coweight lattice (which equals $N_{aff} \cdot o$)

2. The number k is the least natural number with this property.

The numbers k_R and k_w are listed in the table (7.3) below (the number i in the table is the index of connection). We will verify the computation in the most interesting case, namely for the root system of the A-type.

LEMMA 7.7. *Suppose that the Dynkin diagram Γ has type A_n and that $F \cong \mathbb{Z}/m$. Then $k = k(W_{aff}, L)$ equals m. In particular, if $L = N_{aff}$ then we get $k_w = n + 1$.*

Proof: The group $F \cong \mathbb{Z}/m$ acts on the graph $\tilde{\Gamma} = \tilde{A}_n$ by cyclic permutations. Let $g \in F$ be a permutation of order t; note that t divides m. Then for each orbit $\mathcal{O}$ of g (in the vertex set of the Weyl alcove a) we get:

$$b(\mathcal{O}) = \frac{1}{t} \sum_{x_j \in \mathcal{O}} x_j.$$

For each $i \neq 0$, $\alpha_i(b(\mathcal{O})) = 0$ if $x_i \notin \mathcal{O}$, and $\alpha_i(b(\mathcal{O})) = 1/t$ if $x_i \in \mathcal{O}$. For the highest root we get: $\theta(b(\mathcal{O})) = \frac{t-1}{t}$ if $x_0 \in \mathcal{O}$ and $\theta(b(\mathcal{O})) = 1$ if $x_0 \notin \mathcal{O}$. In any case, $k_{\mathcal{O}} = t$.

Since all t's divide the order m of the group F (and for the generator of F, $t = m$), the LCM of $k_{\mathcal{O}}$'s taken over all orbits and all elements of F, equals m. □

Similarly we have

LEMMA 7.8. *Suppose that the Dynkin diagram Γ has type D_ℓ and $F \cong \mathbb{Z}/2$. Then $k = k(W_{aff}, L)$ equals 4 if F permutes at least two roots labelled by 2 and $k = 2$ if it does not (the latter holds for the orthogonal groups).*

We note that for all classical root systems except D_4 where $k_w = k_R$, k_w equals the index of connection i and for all exceptional root systems, $k_R = k_w$ (so for the computation of saturation constants for Problem **Q3** D_4 behaves like an exceptional root system).

If $\underline{G}$ is a reductive algebraic group with the cocharacter lattice L and the associated affine Weyl group W_{aff}, we let $K(G) = k_{inv\ fact}(\underline{G}) := k(W_{aff}, L)$. The numbers $k_{inv\ fact}$ listed below appear in the discussion of the saturation for the invariant factor problem for the group $\underline{G}$, see section 7.

We then obtain:

(7.3)

Root system	$\underline{G}$	θ	i	k_R	k_w	$k(\underline{G})$
A_ℓ	$SL(\ell+1)$	$\alpha_1 + ... + \alpha_\ell$	$\ell+1$	1	$\ell+1$	1
A_ℓ	$GL(\ell+1)$					1
A_ℓ	$PSL(\ell+1)$					$\ell+1$
B_ℓ	$Spin(2\ell+1)$, $SO(2\ell+1)$	$\alpha_1 + 2\alpha_2 + ... + 2\alpha_\ell$	2	2	2	2
C_ℓ	$Sp(2\ell)$, $PSp(2\ell)$	$2\alpha_1 + 2\alpha_2 + ...$ $+2\alpha_{\ell-1} + \alpha_\ell$	2	2	2	2
D_ℓ	$Spin(2\ell)$, $SO(2\ell)$	$\alpha_1 + \alpha_2 + \alpha_3 +$ $+2\alpha_4 + ... + 2\alpha_\ell$	4	2	4	2
D_ℓ, $\ell > 4$	$PSO(2\ell)$					4
D_4	$PSO(8)$	$\alpha_1 + \alpha_2 + \alpha_3 + 2\alpha_4$	4	2	2	2
G_2	G	$3\alpha_1 + 2\alpha_2$	1	6	6	6
F_4	G	$2\alpha_1 + 3\alpha_2 + 4\alpha_3 + 2\alpha_4$	1	12	12	12
E_6	$\tilde{G}$, $Ad(G)$	$\alpha_1 + \alpha_2 + 2\alpha_3$ $+2\alpha_4 + 2\alpha_5 + 3\alpha_6$	3	6	6	6
E_7	$\tilde{G}$, $Ad(G)$	$\alpha_1 + 2\alpha_2 + 2\alpha_3 + 2\alpha_4$ $+3\alpha_5 + 3\alpha_6 + 4\alpha_7$	2	12	12	12
E_8	G	$2\alpha_1 + 2\alpha_2 + 3\alpha_3 + 3\alpha_4 +$ $4\alpha_5 + 4\alpha_6 + 5\alpha_7 + 6\alpha_8$	1	60	60	60

Here $\tilde{G}$ denotes the simply-connected algebraic group, the symbol $Ad(G)$ denotes the algebraic group of adjoint type, i.e. the quotient of $\tilde{G}$ by its center. In the case of root systems with the index of connection equal to 1, $Ad(G) = \tilde{G}$, so we have used the symbol G to denote the unique connected algebraic group with the given root system. Note that for the non-simply-connected classical groups we always get the order of the fundamental group as the saturation factor (except for the group $PSO(8)$).

7.2. The existence of fixed vertices

We begin this section with few simple remarks about existence of polygons in X with the vertices of a given type. As before, $G \subset Aut(X)$ is such that the G-stabilizer of the model apartment A acts on A through the group $\widetilde{W}$, whose translational part is $L = L_{\underline{G}}$. Thus, if $\overline{xy} \subset X$ is a geodesic segment with $x \in G \cdot o$, then $y \in G \cdot o$ iff $\sigma(x, y) \in L$. Therefore the following are equivalent for a polygon $P \subset X$, whose Δ-lengths are in $L = L \cdot o$:

1. All vertices of P are in $G \cdot o$.
2. One (say, the first) of the vertices of P is in $G \cdot o$.

We also have

LEMMA 7.9. *For a vector* $\vec{\tau} \in D_n(X) \cap L^n$ *the following are equivalent:*

1. There exists a polygon $P \subset X$ *with the* Δ*-lengths* $\vec{\tau}$*, such that all (equivalently, one of) the vertices of* P *are in* $G \cdot o$*.*

2. There exists a polygon $P \subset X$ *with the* Δ*-lengths* $\vec{\tau}$*, such that all (equivalently, one of) the vertices of* P *are special.*

Proof: It is clear that (1)⇒(2), let's prove the converse. Let $P = x_1 \cdots x_n$ be a polygon in X with the Δ-lengths $\vec{\tau}$, such that the first vertex x_1 of P is special. Let $y_1, ..., y_{n+1} \in E$ be such that $\overline{y_i y_{i+1}} \subset E$ have the same refined lengths as the segments $\overline{x_i x_{i+1}}$, $i = 1, ..., n$. By assumption, the vertex y_1 is special. Recall that the normalizer N_{aff} of W_{aff} in the group of translations of E, acts transitively on the set of special vertices. We identify the model apartment E with an apartment in X containing the origin o. Let $T \in N_{aff}$ be such that $T(y_1) = o$. Set $z_i := T(y_i)$, $i = 1, ..., n$. Then z_i are in L-orbit of o for each i and according to the Transfer Theorem 4.1, there exists a geodesic polygon $Q = u_1 \cdots u_n \subset X$ with the refined side-lengths same as for $\overline{z_i z_{i+1}}, i = 1, ..., n$. In particular, $\sigma(u_i, u_{i+1}) = \tau_i$ for each i and all vertices of Q are in $G \cdot o$. □

Thus a vector $\vec{\tau}$ belongs to the image of the map (7.1) iff there exists a polygon $P \subset X$ with special vertices and the Δ-side lengths $\vec{\tau}$.

As we have seeing in chapter 5, the existence problem for polygons with the given Δ-side lengths in X is equivalent to the existence of a fixed point for a certain map $\Phi_\psi : X \to X$. In this section we will try to find conditions under which Φ_ψ fixes a *special vertex* in the building X. We first analyze the case when the Euclidean Coxeter complex is irreducible.

We therefore assume that the affine Coxeter group W_{aff} corresponds to a reduced irreducible root system R of rank ℓ in an ℓ-dimensional vector space V, with the underlying affine space E, see chapter 2. Recall that L is a lattice such that

$$L_{trans} = Q(R^\vee) \subset L \subset N_{aff} = P(R^\vee). \tag{7.4}$$

As before, $\tilde{W}$ denotes the subgroup of $\mathrm{Isom}(E)$ generated by elements of W_{aff} and L. Then $\tilde{W}$ is a Euclidean Coxeter group with the linear part W_{sph} and translation part L. Since $\tilde{W}$ normalizes W_{aff}, we get the induced action of the finite abelian group $\tilde{W}/W_{aff} \cong L/L_{trans} = F$ on the Weyl alcove α of W_{aff}. The notion of L-admissible faces of Weyl alcoves of (E, W_{aff}) introduced in section 7.1, carries over to the building X. Note that in the case $L = L_{trans}$, the only L-admissible faces are the vertices of X.

Suppose that $\tau = (\tau_1, ..., \tau_n)$, where $\tau_i \in L$ for each i. Let

$$\psi = ((m_1, \xi_1), ..., (m_n, \xi_n))$$

be a weighted configuration on $\partial_{Tits} X$ of the type τ, where $m_i = \tau_i/|\tau_i|$. Consider the map $\Phi = \Phi_\psi : X \to X$. Then $\Phi = \Phi_n \circ ... \circ \Phi_1 : X \to X$ where each Φ_i acts on every apartment a_i asymptotic to ξ_i by the translation T_i (by the vector which has the same Δ-length as τ_i). Thus the assumption $\tau_i \in L \subset N_{aff}$ implies that each T_i is an automorphism of the Euclidean Coxeter complex (which we identify with a_i). In particular, the map Φ preserves the simplicial structure of X.

THEOREM 7.1. *Assume in addition that Φ has a fixed point in X. Then:*
1. Φ fixes the barycenter of an L-admissible face in X.
2. If $\sum_{i=1}^n \tau_i \in L_{trans}$ then Φ fixes a vertex of X.
3. More generally, if $\hat{L}$ is a lattice normalized by W_{sph}, such that $L_{trans} \subset \hat{L} \subset L$, and $\sum_{i=1}^n \tau_i \in \hat{L}$, then Φ fixes the barycenter of an $\hat{L}$-admissible face in X.
4. If Φ fixes a special vertex of X then $\sum_{i=1}^n \tau_i \in L_{trans}$. In particular, if the Coxeter complex associated with X is the irreducible Coxeter complex (E, W_{aff}) of type A_ℓ, we have:

$$\sum_{i=1}^n \tau_i \in L_{trans} \iff \Phi \textit{ fixes a vertex of } X.$$

Proof: 1. For a fixed point $x = x_1$ of Φ, let δ denote the smallest face (a simplex) in X containing x. Consider the apartments $a_1 \subset X$ (containing x_1, x_2, ξ_1), $a_2 \subset X$ (containing x_2, x_3, ξ_2),..., $a_n \subset X$ (containing x_n, x_1, ξ_n). Then $\Phi_1(\delta) \subset a_1 \cap a_2$, $\Phi_2 \circ \Phi_1(\delta) \subset a_2 \cap a_3$,..., $\Phi_n \circ ... \circ \Phi_1(\delta) \subset a_n \cap a_1$. Since Φ preserves the simplicial structure of X and δ is the smallest simplex containing x, we get: $\Phi(\delta) = \delta$.

The map $\Phi|_\delta : \delta \to \delta$ is the composition of maps

$$(T_n \circ ... \circ T_2 \circ T_1)|_\delta,$$

where each $T_i : a_i \to a_i$ is a translation. Let $\varphi_1 : E \to a_1$ be an (isometric) parameterization as in the definition of a space modeled on a Coxeter complex. We can assume that $\delta' := \varphi_1^{-1}(\delta)$ is a face of the Weyl alcove $\alpha \subset E$. There exists a parameterization $\varphi_2 : E \to a_2$ so that $\varphi_2^{-1} \circ \varphi_1 = id$ on the domain of this composition. Similarly (like in the definition of a developing map of a geometric structure) we choose parameterizations $\varphi_i, i = 1, ..., n$, so that for each i, $\varphi_{i+1}^{-1} \circ \varphi_i = id$ on the domain of this composition. Then we get:

$$(T_n \circ ... \circ T_2 \circ T_1)|_\delta = (T_n \circ \varphi_n \circ \varphi_{n-1}^{-1} \circ T_{n-1} \circ ... \circ \varphi_2^{-1} \circ T_2 \circ \varphi_2 \circ \varphi_1^{-1} \circ T_1)|_\delta.$$

Set

$$T_i' := \varphi_i^{-1} \circ T_i \circ \varphi_i : E \to E.$$

These maps are translations by the vectors τ_i', which are in the W_{aff}-orbits of the vectors τ_i, $i = 1, ..., n$. Set

$$T' := T_n' \circ ... \circ T_1'.$$

Then the "holonomy map"

$$T' = \varphi_n^{-1} \circ T_n \circ \varphi_n \circ \varphi_{n-1}^{-1} \circ ... \circ \varphi_2^{-1} \circ T_2 \circ \varphi_2 \circ \varphi_1^{-1} \circ T_1 \circ \varphi_1 : E \to E$$

sends the simplex δ' to itself. Moreover,

$$T'|_{\delta'} = \varphi_n^{-1} \circ \Phi \circ \varphi_1|_{\delta'} = \varphi_n^{-1} \circ \varphi_1 \circ \varphi_1^{-1} \circ \Phi \circ \varphi_1|_{\delta'} = w \circ \varphi_1^{-1} \circ \Phi \circ \varphi_1|_{\delta'},$$

where $w \in W_{aff}$. Thus, by letting

$$\Phi'|_{\delta'} := \varphi_1^{-1} \circ \Phi \circ \varphi_1|_{\delta'},$$

we conclude that $\Phi'|_{\delta'}$ admits the extension by $w^{-1}T'$ to the entire E. Since $w^{-1}T' \in \tilde{W}$, Φ induces an automorphism of the simplex δ by an element of the group $\tilde{W}$. We note that Φ does not necessarily permute all the vertices of δ. Consider however an orbit of $\langle \Phi \rangle$ on the vertex set of δ. It is clear that this orbit spans an L-admissible simplex c in $\delta \subset X$; since $\Phi|_\delta$ is a 1-Lipschitz automorphism, it preserves the barycenter of the simplex c and hence the first assertion of the theorem follows.

2. Let us prove the second assertion. Set $t = \sum_{i=1}^n \tau_i$. Note that the map T_i' is the translation in E by the vector τ_i' where $\tau_i' = w_i(\tau_i)$ for some $w_i \in W_{sph}$. The composition $T' = T_n' \circ ... \circ T_1'$ is the translation by the vector

$$v = \sum_{i=1}^n \tau_i' = \sum_{i=1}^n w_i(\tau_i).$$

We claim that $v \in L_{trans}$ iff $t \in L_{trans}$. We leave the proof of the following elementary lemma to the reader:

LEMMA 7.10. *If $\tau \in P(R^\vee)$ then for each $w \in W_{sph}$ we have:*

$$w(\tau) - \tau \in Q(R^\vee).$$

Therefore

$$v = \sum_{i=1}^n (w_i(\tau_i) - \tau_i) + \sum_{i=1}^n \tau_i = s + t,$$

where $s \in Q(R^\vee) = L_{trans}$. This proves the claim.

Recall that $\Phi'|_{\delta'}$ admits the extension by $w^{-1}T'$ (where $w \in W_{aff}$) to the entire model apartment E. Thus, if $t \in L_{trans}$ then $\Phi'|_{\delta'}$ is the restriction of an element $g \in W_{aff}$. Since the alcove α is a fundamental domain for the action of W_{aff} on E and δ' is a face of α, we conclude that $\Phi'|_{\delta'} = id$. Hence Φ fixes a vertex of X.

3. The proof of 3 is analogous to the proof of 2 and we leave it to the reader.

4. Lastly, suppose that Φ fixes a special vertex x of X, then $\delta = \{x\}$, $\delta' = \{x'\}$ is a special vertex in E, and $\Phi'(x') = x'$ implies that $T'(x') = w(x')$, where $w \in W_{aff}$. Since x' is a special vertex, $w = T \circ w'$, where $w' \in W_{aff}$ fixes x' and $T \in L_{trans}$. Thus $T'(x') = T(x')$, which implies that $T' \in L_{trans}$, and hence $t \in L_{trans}$. □

As an immediate corollary of the above theorem we get:

COROLLARY 7.11. *1. For each* $\vec{\tau} = (\tau_1, ..., \tau_n) \in D_n(X) \cap L^n$ *such that* $\sum_i \tau_i \in L_{trans}$*, there exists a polygon* $P \subset X$ *such that* $\sigma(P) = \vec{\tau}$ *and the vertices of* P *are at the vertices of* X.

2. If there exists a polygon $P \subset X$ *with special vertices and* $\sigma(P) = \vec{\tau}$*, then* $\sum_i \tau_i \in L_{trans}$.

EXAMPLE 7.12. For the irreducible Coxeter complex (E, W_{aff}) of the type C_2, there are vectors $\tau_1, \tau_2, \tau_3 \in L = P(R^\vee)$ such that Φ fixes a vertex of X but $\sum_{i=1}^3 \tau_i \notin Q(R^\vee) = L_{trans}$.

Proof: Consider the billiard triangle T shown in Figure 1. It has geodesic sides $\overline{zx}$, $\overline{zy}$ and the broken side $\overline{xuy}$. Its Δ-side lengths are L_{trans}-integral and it has non-special vertices.

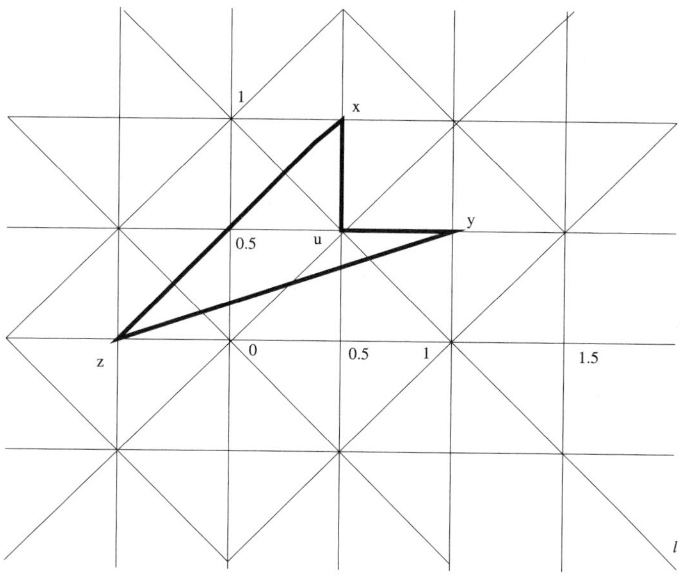

FIGURE 1. *A billiard triangle.*

Let us check that this triangle can be unfolded to a geodesic triangle $\Delta(z, x, y')$ in X (with vertices at the vertices of X), in the sense of section 6.1. This can be done as follows. We note that the geodesic l (weakly) separates z from x and y. Hence we can unfold the billiard triangle T to a geodesic triangle in X using Lemma 6.7.

Then the vectors τ_1, τ_2 representing the sides $\overline{xz}$ and $\overline{xy'}$ of $\Delta(x, y', z)$ are in the coroot lattice $Q(R^\vee)$, but the vector τ_3 representing the side $\overline{zy}$ is not in the coroot lattice. Hence $\sum_{i=1}^3 \tau_i \notin Q(R^\vee)$. □

Thus the fixed point of Φ may not be special. However, if N_{aff} acts transitively on the vertices of E, every vertex is special and we get:

COROLLARY 7.13. *Suppose that the normalizer* N_{aff} *of* W_{aff} *in* V *acts transitively on the vertices of* E. *Then the image of* $D_n^{ref,L}(X)$ *in* $D_n(X)$ *equals*

$$D_n^{L,0}(X) := \{(\tau_1, ..., \tau_n) \in D_n(X) \cap L^n, \sum_{i=1}^n \tau_i \in L_{trans}\}.$$

The example of a Coxeter complex for which the hypothesis of Corollary 7.13 holds is given by (E, W_{aff}) for which the finite Weyl group W_{sph} is of type A_ℓ. Then each vertex of the Coxeter complex (E, W_{aff}) is special.

Next we consider the general case when the Coxeter complex (E, W_{aff}) is reducible. As in section 4.2, we have the corresponding deRham decomposition of the building X:

$$X = X_0 \times X_1 \times ... \times X_s,$$

where X_0 is the flat deRham factor, the group W^i_{aff} $(i > 0)$ acts as an irreducible affine Coxeter group on the Euclidean space E_i. The Weyl chamber Δ is the product $\prod_{i=0}^{s} \Delta_i$ of Weyl chambers, where of course $\Delta_0 = E_0$. Let p_i denote the orthogonal projection $X \to X_i$, $i = 0, 1, ..., s$; by abusing notation we will also use p_i to denote the orthogonal projections $E \to E_i$. Given a lattice L' in E we get the inclusions

$$L^i_{trans} \subset p_i(L) = L_i \subset N^i_{aff}, i = 1, ..., s.$$

Note that

$$D_n(X_0) = \{(\tau_1, ..., \tau_n) : \sum_{i=1}^{n} \tau_i = 0\}.$$

We have

PROPOSITION 7.14. *1.* $D_n(X) = \prod_{i=0}^{s} D_n(X_i)$.

2. For each $\vec{\tau} \in \Delta^n$, *there is a polygon* $P \subset X$ *with the* Δ*-side lengths* $\vec{\tau}$ *and the vertices at the (special) vertices of* X *iff for each* i *there exists a polygon* $P_i \subset X_i$ *with the* Δ_i*-side lengths* $p_i(\vec{\tau})$ *and the vertices at the (special) vertices of* X_i.

3. $D_n^{ref,L'}(X) = \prod_{i=0}^{s} D_n^{ref,L'_i}(X_i)$.

Proof: We will prove the third assertion since the proofs of (1) and (2) are similar. First of all, if $P \subset X$ is a polygon with L'-integral refined side-lengths then its projections $P_i := p_i(P) \subset X_i$ are also polygons with L_i-integral refined side-lengths. Conversely, if $P_i \subset X_i$ are polygons with L'_i-integral refined side-lengths and x_{ij} is the j-th vertex of the polygon P_i, we set

$$x_j := (x_{0j}, ..., x_{sj}) \in X,$$

this point is a vertex of X since all x_{ij}'s are vertices of the corresponding buildings X_i. This defines the polygon $P = x_1 \cdots x_n$ in X. It is clear that the refined side-lengths of P are L'-integral. □

The above proposition implies that Corollaries 7.11 and 7.13 remain valid for Euclidean buildings with reducible Coxeter complexes.

As in section 7.1, given a lattice L' in E we define the lattice L as the intersection of L' and the translation group of $E_1 \oplus ... \oplus E_s$. We now can compute the image of the map (7.1) "up to saturation":

THEOREM 7.2. *1. Let* $k = k(W_{aff}, L') = k(W_{aff}, L)$ *be the saturation factor defined in section 7.1. Then for each* $\vec{\tau} \in D_n(X) \cap (kL')^n$, *there exists an* n*-gon* $P \subset X$ *with special vertices and* Δ*-side lengths* $\tau_1, ..., \tau_n$.

2. Let R *be the root system corresponding to the Coxeter complex* (E, W_{aff}) *and let* $k := k_R = k(W_{aff}, L_{trans})$ *be the saturation factor defined in section 7.1.*

Then for each $\vec{\tau} \in D_n(X) \cap (kL')^n$ *satisfying*

$$\sum_i \tau_i \in L_{trans} = Q(R^\vee)$$

there exists an n*-gon* $P \subset X$ *with special vertices and* Δ*-side lengths* $\tau_1, ..., \tau_n$.

Proof: 1. We first consider the case when the root system R is reduced, irreducible and has rank equal the dimension of the space V; then $L = L'$. Let $P' \subset X$ be a polygon with the Δ-lengths $\frac{1}{k}\tau_1, ..., \frac{1}{k}\tau_n$. Since $\frac{1}{k}\tau_i \in L$ for each i, Theorem 7.1 implies that the polygon P' can be chosen so that its first vertex is the barycenter of an L-admissible simplex in X. Let $\overline{x_i x_{i+1}}$, $i = 1, 2, ..., n$, be oriented segments in the model apartment E, which represent the refined side-lengths of the polygon P'; the point x_1 (and hence x_i for each $i = 2, ..., n+1$) is the barycenter of an L-admissible simplex in E. We regard E as the vector space V by identifying the origin with the special vertex o. Then, according to the definition of $k = k(W_{aff}, L)$, the segments $\overline{(kx_i)(kx_{i+1})} = \overline{y_i y_{i+1}}$ have end-points at the *special vertices* of E. Consider the affine transformation $\iota : E \to E$ which is the homothety $x \mapsto kx$. Then $\iota : (E, W_{aff}) \to (E, W_{aff})$ is an injective endomorphism of the affine Coxeter systems and the Transfer Theorem 4.1 implies that there exists a polygon P in X whose refined side-lengths are represented by $\overline{y_i y_{i+1}}$, $i = 1, ..., n$. The vertices of P are special and its Δ-side-lengths are $\tau_1, ..., \tau_n$.

We now consider the general case when the Coxeter complex (E, W_{aff}) is reducible. Let $P' \subset X$ be a polygon with the Δ-lengths $\frac{1}{k}\tau_1, ..., \frac{1}{k}\tau_n$. Since

$$p_0(\sum_{i=1}^n \tau_i) = 0,$$

we have

$$\frac{1}{k}\sum_{i=1}^n \tau_i \in L' \cap V_1 \oplus ... \oplus V_s = L.$$

Hence for each $j = 1, ..., s$ we get

$$\frac{1}{k}\sum_{i=1}^n p_j(\tau_i) \in L_j = p_j(L).$$

Therefore, by Theorem 7.1 (part 3, where we take $\hat{L} = L_j$, $j = 1, ..., s$), the polygon P' can be chosen so that for each $j = 1, ..., s$, the vertices of $p_j(P')$ are barycenters of L_j-admissible faces of X_j. Recall that

$$k = LCM(k_1, ..., k_s).$$

By the irreducible case, for each $j = 1, ..., s$ there is a polygon $P'_j \subset X_j$ with special vertices and side-lengths

$$\frac{k_j}{k} p_j(\tau_1), ..., p_j(\tau_n).$$

Let $\overline{x_{i,j} x_{i+1,j}}$, $i = 1, 2, ..., n, j = 1, ..., s$, be oriented segments in the model apartment E_j, which represent the refined side-lengths of the polygon P'_j. It follows from the Transfer Theorem 4.1 that for each $j = 1, ..., s$ there exists a polygon $P_j \subset X_j$

whose refined side-lengths are represented by the segments $\overline{(m_j x_{i,j})(m_j x_{i+1,j})}$, where $m_j = k/k_j$. Note that

$$m_j \frac{k_j}{k} p_j(\tau_i) = p_j(\tau_i)$$

for each i and j. By the definition of k_j, for each j the vertices of P_j are special vertices of X_j. Now Proposition 7.14 implies that there exists a polygon $P \subset X$ with the special vertices, whose projections to X_j are P_j for each j and the Δ-side lengths of P are $\tau_1, ..., \tau_n$.

2. The proof of this assertion is analogous to the proof of 1. First, suppose that the Coxeter complex is irreducible. Then, according to Theorem 7.1 (part 2), a polygon $P' \subset X$ with the Δ-lengths $\frac{1}{k}\tau_1, ..., \frac{1}{k}\tau_n$ can be chosen so that its vertices are vertices of X. Let $\overline{x_i x_{i+1}}$, $i = 1, 2, ..., n$, be oriented segments in the model apartment E, which represent the refined side-lengths of the polygon P'. Then, by the definition of $k = k_R$, the end-points of the segments $\overline{(kx_i)(kx_{i+1})}$ are special vertices of X. The rest of the argument is the same as for (1). □

7.3. Saturation factors for reductive groups

We now apply the results from the previous section to the generalized Invariant Factor Problem for nonarchimedean reductive Lie groups G. Suppose that $\underline{G}$ is a connected reductive algebraic Lie group over $\mathbb{K}$, where $\mathbb{K}$ is a field with discrete valuation v. Let $G := \underline{G}(\mathbb{K})$. The group $\underline{G}$ determines a Bruhat-Tits building X, Bruhat-Tits root system R, the corresponding affine Coxeter group W_{aff} and the extended cocharacter lattice $L := L_{\underline{G}}$ satisfying the double inclusion (7.4). Define $k_{inv\ fact}(G) := k(W_{aff}, L)$.

As an immediate corollary of Theorem 7.2, we obtain:

COROLLARY 7.15. *1. For $k = k_{inv\ fact}(G)$ and any $\tau \in L^n \cap D_n(X)$, there exists an n-gon P in X with Δ-side-lengths equal to $k\tau$ and vertices in $G \cdot o$.*

2. For $k = k_R$ and any

$$\vec{\tau} \in D_n^{L,0}(X) = \{\vec{\tau} \in D_n(X) \cap L^n : \sum_{i=1}^{n} \tau_i \in Q(R^\vee)\},$$

there exist an n-gon P with Δ-side lengths equal to $k\tau$ and vertices in $G \cdot o$.

3. If $\underline{G}$ is semisimple and simply-connected then $D_n^{L,0}(X) = D_n^L(X)$ and $k_{inv\ fact}(G) = k_R$.

Proof: Parts (1) and (2) follow from Theorem 7.2 (parts (1) and (2) respectively) and Lemma 7.9.

To prove (3) note that in this case $Q(R^\vee) = L$ and thus

$$k_{inv\ fact}(G) = k_R = k(W_{aff}, Q(R^\vee)). \quad \square$$

REMARK 7.16. We emphasize that $k_{inv\ fact}(G)$ and k_R depend only on G and not the choice of τ.

By specializing to the case $n = 3$ and using the equivalence (see §4.4) of Problem RGTI for Euclidean buildings with the Problem **Q3**, we get:

COROLLARY 7.17. *For each $k \in \mathbb{N}$ divisible by $k_{inv\ fact}(G)$ and each triple $(\alpha, \beta, \gamma) \in k(L \cap \Delta)^3$ which satisfies the* generalized triangle inequalities *for the space X, there exist representatives A, B, C of the double coset classes*

$$\underline{G}(\mathcal{O})\alpha(\pi)\underline{G}(\mathcal{O}), \quad \underline{G}(\mathcal{O})\beta(\pi)\underline{G}(\mathcal{O}), \quad \underline{G}(\mathcal{O})\gamma(\pi)\underline{G}(\mathcal{O}),$$

such that $ABC = 1$.

EXAMPLE 7.18. Let $\underline{G} = Spin(5)$. Then Example 6.9 shows that there is a triple $(\alpha, \beta, \gamma) \in (L \cap \Delta)^3, L = Q(R^\vee)$, such that α, β, γ satisfy the generalized triangle inequalities and such that one cannot find A, B, C in the corresponding double cosets of $Spin(5, \mathbb{K})$ such that $ABC = 1$.

Thus one cannot take $k = 1$ for the case $\underline{G} = Spin(5)$.

EXAMPLE 7.19. $k_{inv\ fact}(GL(m)) = 1$. In other words, let X be the Bruhat-Tits building associated with the group $GL(m, \mathbb{K})$, where $\mathbb{K}$ is a field with nonarchimedean discrete valuation and the value group $\mathbb{Z}$. Let $\vec{\tau} \in \mathbb{Z}^m$ have entries arranged in decreasing order. Then $\vec{\tau}$ satisfies the generalized triangle inequalities (i.e. $\vec{\tau} \in D_n(X)$) if and only if there exists an n-gon P in X with the first vertex at the origin (the point stabilized by $GL(m, \mathcal{O})$) and Δ-side lengths $\vec{\tau}$.

Although this example is a special case of Theorem 7.2 (part 2), we present a complete proof for the benefit of the reader.

Proof: Recall (see Example 7.3) that the Euclidean Coxeter complex (E, W_{aff}) of X is reducible, $E = E_0 \oplus E_1$, where E_1 is the kernel of the map

$$tr : (x_1, ..., x_m) \mapsto x_1 + ... + x_m,$$

and E_0 is the span of $(1, ..., 1)$. Let V_1 be the vector space underlying E_1. We may identify V_1 with the Lie algebra of traceless real diagonal matrices, the real Cartan subalgebra of the Lie algebra of $SL(m)$. Moreover (see Example 7.3), $L = L_{GL(m)} = \mathbb{Z}^m$, is the cocharacter lattice; the orthogonal projection of L to V_1 is the coweight lattice of $SL(m)$, and the root system is of type A_{m-1}. On the other hand, suppose that $\vec{\tau} = (\tau_1, ..., \tau_n)$. Then, since $\vec{\tau} \in D_n(X)$, there exists a polygon $\hat{P}$ in X with the Δ-side lengths τ_i. Let $X = X_0 \times X_1$ be the deRham decomposition of X, where $X_0 = E_0 \cong \mathbb{R}$ is the flat factor. Since the projection of $\hat{P}$ to the flat deRham factor E_0 of X is also a polygon, we get:

$$p_0(\tau_1 + ... + \tau_n) = tr(\tau_1 + ... + \tau_n) = 0.$$

Hence $\tau_1 + ... + \tau_n \in L \cap E_1 = Q(R^\vee)$. Therefore we can apply Theorem 7.1 (part 2 or 4) to conclude that there exists a polygon $P' \subset X_1$ whose vertices are (special) vertices of X_1 and whose Δ-side lengths are $p_1(\tau_1), ..., p_1(\tau_n)$. By combining this with the polygon $p_0(\hat{P})$ via Proposition 7.14 we conclude that there exists a polygon $P \subset X$ whose vertices are (special) vertices of X and whose Δ-side lengths are $\tau_1, ..., \tau_n$. Thus $k_{inv\ fact}(GL(m)) = 1$. □

THEOREM 7.3. *1. Let $\underline{DG}$ be the derived (i.e. commutator) subgroup of $\underline{G}$. Then*

$$k_{inv\ fact}(\underline{G}) = k_{inv\ fact}(\underline{DG}). \tag{7.5}$$

2. Moreover, if $\underline{G_1}, ..., \underline{G_s}$ are the simple factors of $\underline{DG}$ then

$$k_{inv\ fact}(\underline{G}) = LCM(k_{inv\ fact}(\underline{G_1}), ..., k_{inv\ fact}(\underline{G_s})).$$

Proof: The equation (7.5) follows from three observations:

We first claim that the lattice $L' = L_{\underline{G}} \cap V_1 \oplus ... \oplus V_s$ equals $L_{\underline{DG}}$. This is because we can assume that the maximal split tori $\underline{T}_{\underline{DG}}$ and $\underline{T}_{\underline{G}}$ of $\underline{DG}$ and $\underline{G}$ are related by

$$\underline{T}_{\underline{DG}} = \underline{T}_{\underline{G}} \cap \underline{DG}.$$

Hence the cocharacter lattice of $\underline{DG}$ is the sublattice of $L_{\underline{G}}$ consisting of those cocharacters whose image is contained in $\underline{DG}$. Passing to tangent vectors at the identity yields the claim.

Second, by definition, the coroot system $R^\vee_{\underline{DG}}$ of $\underline{DG}$ is the same as the coroot system $R^\vee_{\underline{G}}$ of $\underline{G}$ regarded as a root system in $E_1 \oplus ... \oplus E_s$. Thus $\underline{G}$ and $\underline{DG}$ have the same affine Weyl group W_{aff}.

Lastly, the projection of the lattice $L_{\underline{G}}$ into $V_1 \oplus ... \oplus V_s$ is contained in $P(R^\vee) \cap V_1 \oplus ... \oplus V_s$.

With these observations, the same argument as in the case of $GL(m)$ goes through:

$$k(W_{aff}, L_{\underline{G}}) = k(W_{aff}, L_{\underline{DG}})$$

(the saturation factor $k_{inv\ fact}$ depends only on the group W_{aff} and the cocharacter lattice), which implies (7.5).

To prove the second assertion of the Theorem note that the Euclidean Coxeter complex for $\underline{DG}$ is the direct product of the Euclidean Coxeter complexes for its factors $\underline{G}_i, i = 1, ..., s$. Let $\underline{T}_i$ be maximal split tori of $\underline{G}_i$, $i = 1, ..., s$ and $L_i = X_*(\underline{T}_i)$ be their cocharacter lattices. Let $\underline{T}_{\underline{G}} := \prod_i \underline{T}_i$, and $L' := X_*(\underline{T}_{\underline{G}})$; then each L_i is the projection of L' to V_i, where (V_i, W^i_{aff}) are the Euclidean Coxeter complexes associated with $\underline{G}_i$.

Then Definition 7.5 implies that

$$k(W_{aff}, L') = LCM(k(W^1_{aff}, L_1), ..., k(W^s_{aff}, L_s)). \qquad \square$$

We can also give a complete solution (in terms of the stability inequalities) for the Invariant Factor Problem **Q3** in the case of nonarchimedean reductive Lie groups G with root system of type A_{m-1}. For instance, $\underline{G}$ could be the quotient of $SL(m)$ by a subgroup of its center. As before, let L denote the lattice $L_{\underline{G}}$ and let $X := X_{\underline{G}}$ be the Bruhat-Tits building associated with $\underline{G}$.

THEOREM 7.4. *Let $\underline{G}$ be a reductive algebraic group over $\mathbb{K}$, so that the group G has the associated root system R of type A_{m-1}. Then the natural embedding*

$$D_3^{ref,L}(X) \hookrightarrow D_3^L(X) = \{(\alpha, \beta, \gamma) \in D_3(X) \cap L^3 : \alpha + \beta + \gamma \in Q(R^\vee)\} \tag{7.6}$$

is onto.

Proof: The assertion follows from Corollary 7.15 and the fact that $k_R = 1$ for the root system of type A. $\square$

Specializing to the case of $GL(m)$ and $SL(m)$ we get the following corollary:

COROLLARY 7.20. *Let $X = \mathbb{P}_m$ be the symmetric space of symmetric positive-definite $m \times m$-matrices. Then there exists a solution to the Invariant Factor Problem* **Q3** *for the case $G = SL(m, \mathbb{K})$ (or $GL(m, \mathbb{K})$) and $K = SL(m, \mathcal{O})$*

(resp. $GL(m, \mathcal{O})$) if and only if α, β and γ are integer diagonal matrices so that $tr(\alpha + \beta + \gamma) = 0$ and the projections

$$\alpha - \frac{1}{m}tr(\alpha)I, \beta - \frac{1}{m}tr(\beta)I, \gamma - \frac{1}{m}tr(\gamma)I$$

satisfy the stability inequalities 5.6 for the space X. Here I is the identity matrix.

CHAPTER 8

The Comparison of Problems Q3 and Q4

8.1. The Hecke ring

In this section we will (for the most part) follow the notation of [**Gro**]. We urge the reader to consult this paper for more details. However we will let $\mathbb{K}$ and $\mathcal{O}$ be as in section 3.3. We will assume that the valuation v is discrete, the field $\mathbb{K}$ is locally compact and the residue field is finite of order q and uniformizer π.

REMARK 8.1. The assumption that $\mathbb{K}$ is locally compact is equivalent to $\mathbb{K}$ being a totally-disconnected local field, see [**Ta**, pg. 5]. By applying Theorem 4.1, it follows that the main result of this section, Theorem 8.5, also holds in the case when $\mathbb{K}$ is not locally compact, for example the case in which $\mathbb{K} = \mathbb{C}((t))$ the field of fractions of the ring of formal power series $\mathbb{C}[[t]]$.

We let $\underline{G}$ be a connected reductive algebraic group over $\mathbb{K}$. We will assume that $\underline{G}$ is split over $\mathbb{K}$. Then $\underline{G}$ is the general fiber of a group scheme (also denoted $\underline{G}$) over $\mathcal{O}$ with reductive special fiber. We fix a maximal split torus $\underline{T} \subset \underline{G}$ defined over $\mathcal{O}$. We put $G = \underline{G}(\mathbb{K}), K = \underline{G}(\mathcal{O})$ and $T = \underline{T}(\mathbb{K})$. We let $\underline{B} \subset \underline{G}$ be a Borel subgroup normalized by $\underline{T}$ and put $B = \underline{B}(\mathbb{K})$. We let U be the unipotent radical of B whence $B = TU$. Let $X = X_{\underline{G}}$ denote the Bruhat-Tits building associated with the group $\underline{G}$; $o \in X$ is a distinguished special vertex stabilized by the compact subgroup K.

We have already defined in section 3.3 the free abelian groups (of rank $l = \dim(\underline{T})$) $X_*(\underline{T})$ and $X^*(\underline{T})$ and a perfect $\mathbb{Z}$-valued pairing $\langle\ ,\ \rangle$ between them. The first contains the coroots $R^\vee$, the second contains the roots R of $\underline{G}$.

The roots are the characters of $\underline{T}$ that occur in the adjoint representation on the Lie algebra of $\underline{G}$. The subset R^+ of the roots that occur in representation on the Lie algebra of $\underline{B}$ forms a positive system and the indecomposable elements of that positive system form a system of simple roots Π. We let W denote the corresponding (finite) Weyl group.

The root basis Π determines a positive Weyl chamber P^+ in $X_*(\underline{T})$, by

$$P^+ = \{\lambda \in X_*(\underline{T}) : \ \langle \lambda, \alpha \rangle \geq 0, \alpha \in \Pi\}.$$

We define a partial ordering on P^+ by $\lambda > \mu$ iff the difference $\lambda - \mu$ is a sum of positive coroots.

We will use the notation 0 for the trivial cocharacter.

We define the element $\rho \in X^*(\underline{T}) \otimes \mathbb{Z}[1/2]$ by

$$2\rho = \sum_{\alpha \in R^+} \alpha.$$

We recall that ρ is the sum of the fundamental weights of R. Then, for all nontrivial $\lambda \in P^+$, the half-integer $\langle \lambda, \rho \rangle$ is positive.

DEFINITION 8.2. The Hecke ring $\mathcal{H} = \mathcal{H}(G, K)$ is the ring of all locally constant, compactly supported functions $f : G \longrightarrow \mathbb{Z}$ which are K-biinvariant. The multiplication in $\mathcal{H}$ is by the convolution

$$f \cdot g(z) = \int_G f(x) \cdot g(x^{-1}z)dx$$

where dx is the Haar measure on G giving K volume 1.

We claim that the function $f \cdot g$ also takes values in $\mathbb{Z}$. Indeed, f and g are finite sums of characteristic functions of K-double cosets. Thus it suffices to prove the claim in the case that f and g are both characteristic functions of K-double cosets. In this case it is immediate that their convolution product is the characteristic function of the set of products of the elements in the two double cosets. This product set is itself a finite union of K-double cosets. This implies that the structure constants $m_{\lambda,\mu}(\nu)$ of the Hecke ring, defined below, are nonnegative integers. The characteristic function of K is the unit element of $\mathcal{H}$. For the proof of the next lemma see [**Gro**, §2].

LEMMA 8.3. *$\mathcal{H}$ is commutative and associative.*

In fact much more is true. For $\lambda \in X_*(\underline{T})$ let c_λ be the characteristic function of the corresponding K-double coset $\lambda(\pi) \in K\backslash G/K$.

LEMMA 8.4. (1) *The assignment $\lambda \longrightarrow c_\lambda$ induces an isomorphism of free abelian groups $\mathbb{Z}[P^+] \longrightarrow \mathcal{H}$.*

(2) *The Hecke ring is a filtered ring with the filtration levels indexed by the ordered abelian semigroup P^+. In particular, we have*

$$c_\lambda \cdot c_\mu = c_{\lambda+\mu} + \sum_{\nu<\lambda+\mu} m_{\lambda,\mu}(\nu)c_\nu. \tag{8.1}$$

Here and in what follows $\sum\limits_{\nu<\mu}$ denotes the sum over elements ν from P^+. We note that this sum is finite.

We will prove in Lemma 8.14, that the structure constants $m_{\lambda,\mu}(\nu)$ are polynomials in q with integer coefficients. Here and below we will keep track of the dependence of certain quantities on the cardinality q of the residue field. Thus q will play the role of a variable in what follows. One of the main points here is that the structure constants $n_{\lambda,\mu}(\nu)$ of the representation ring $R(G^\vee)$ do not depend on q. They will be encoded in the coefficients of the polynomials $m_{\lambda,\mu}(\nu)$ (along with the coefficients of the Kazhdan-Lusztig polynomials $b_\lambda(\mu)$).

We recall the definition of the structure constants $m_{\alpha,\beta,\gamma}(\delta)$: Given double cosets of $\alpha(\pi), \beta(\pi), \gamma(\pi)$ in $K\backslash G/K$, consider the characteristic functions c_α , c_β and c_γ of these double cosets; then decompose the triple product

$$c_\alpha \cdot c_\beta \cdot c_\gamma = \sum_\delta m_{\alpha,\beta,\gamma}(\delta)\ c_\delta$$

in the Hecke algebra $\mathcal{H}$. This defines $m_{\alpha,\beta,\gamma}(\delta)$. Our primary interest is $m_{\alpha,\beta,\gamma}(0)$ (where 0 is the trivial character); it will be viewed as a function of the variable q.

In our proof of the saturation conjecture for $GL(m)$ we will need the following lemma, where γ^* denotes the weight contragredient to γ. Then $\gamma^*(\pi)$ is a representative for the double coset obtained by inverting the elements in the one represented by $\gamma(\pi)$.

LEMMA 8.5.
$$m_{\alpha,\beta,\gamma}(0) = vol(K\gamma(\pi)K)\ m_{\alpha,\beta}(\gamma^*).$$

Proof: We define an inner product $((\ ,\ ))$ on $C_0(G)$, the space of compactly supported complex-valued functions on G, by
$$((f,g)) := \int_G f(x)\overline{g(x)}dx.$$

We first claim that
$$((c_\alpha, c_\beta)) = \begin{cases} vol\ K\alpha(\pi)K, & \text{if } \alpha = \beta \\ 0, & \text{otherwise.} \end{cases}$$

Indeed
$$((c_\alpha, c_\beta)) = \int_G c_\alpha(x)c_\beta(x)dx.$$

The claim follows by noting that the product function is identically one if the cosets coincide and otherwise it is identically zero.

Next we observe that
$$((c_\alpha, c_\beta)) = c_\alpha \cdot c_{\beta^*}(1).$$

Indeed, the right-hand side is equal to
$$\int_G c_\alpha(x)c_{\beta^*}(x^{-1})dx = \int_G c_\alpha(x)c_\beta(x)dx.$$

Hence
$$c_\alpha \cdot c_{\beta^*}(1) = \begin{cases} vol\ K\alpha(\pi)K, & \text{if } \alpha = \beta \\ 0, & \text{otherwise.} \end{cases}$$

Finally
$$m_{\alpha,\beta,\gamma^*}(0) = \sum_\delta m_{\alpha,\beta}(\delta)(c_\delta \cdot c_{\gamma^*})(1) = m_{\alpha,\beta}(\gamma) vol\ K\gamma(\pi)K. \quad \square$$

8.2. A geometric interpretation of $m_{\alpha,\beta,\gamma}(0)$

In this section we will prove the following

THEOREM 8.1. *$m_{\alpha,\beta,\gamma}(0)$ is the number of triangles in the building X with first vertex o and side-lengths α, β, γ.*

We recall that given $\tau \in \Delta$ we define $S(o,\tau) = \{x \in X : \sigma(o,x) = \tau\}$, the "$\Delta$-sphere of radius τ and center at o".

LEMMA 8.6. *For each $\tau \in \Delta$ the group K acts transitively on $S(o,\tau)$.*

Proof: Let $x \in S(o,\tau)$. By the properties of the action of G on the building X, presented in the List 4.10, there exists $g \in G$ such that $g \cdot o = o$ and $g \cdot x \in \Delta \subset A$ where A is the model apartment. Since $g \cdot o = o$ we have $g = k \in K$, whence
$$\overrightarrow{o\ k \cdot x} = \overrightarrow{o\ \tau}, \text{(equality of vectors)}.$$

If follows that $k \cdot x = \tau$ and $x = k^{-1}\tau$. $\square$

As a consequence we can identify the right K–quotients of the K–double cosets in G to the Δ–spheres $S(o,\tau)$ in the building. The proof of the following lemma is then clear.

LEMMA 8.7. *Let $\alpha \in X_*(\underline{T})$. Then the map $k \longrightarrow k \cdot \alpha(\pi)$ induces a bijection between the quotient $K\alpha(\pi)K/K$ and $S(o, \alpha(\pi))$.*

We define

$$f : K\alpha(\pi)K/K \times K\gamma^*(\pi)K/K \longrightarrow K\backslash G/K$$

by sending (g_1, g_2) to the double coset represented by $f(g_1, g_2) := g_1^{-1} \cdot g_2$. The reader will observe that f is well-defined. Thus f induces a map (again denoted f) from $S(o, \alpha(\pi)) \times S(o, \gamma^*(\pi))$ to $K\backslash G/K$.

Now let $\alpha, \beta, \gamma \in X_*(\underline{T})$. We define the set

$$\mathcal{T}_{\alpha,\beta,\gamma} := \{(x, y) \in K\alpha(\pi)K/K \times K\gamma^*(\pi)K/K : f(x, y) \in K\beta(\pi)K\}.$$

Note that this set is finite since $q < \infty$. Let $\Delta(o, x_1, x_2)$ be a triangle in the building X with the Δ-side lengths α, β, γ. Then $f(x_1, x_2) \in K\beta(\pi)K$ and we obtain a map $F : \Delta(o, x_1, x_2) \mapsto (x_1, x_2)$ from the space of triangles in X with side-lengths α, β, γ and first vertex at o into the set $\mathcal{T}_{\alpha,\beta,\gamma}$. We leave the proof of the following lemma to the reader.

LEMMA 8.8. *F is a bijection.*

REMARK 8.9. The appearance of the contragredient coweight γ^* comes about because we require $\sigma(x_2, o) = \gamma$, but $\sigma(x_2, o)$ is the contragredient of the $\sigma(o, x_2)$.

Write

$$K\alpha(\pi)K = \cup_{i=1}^{I} x_i K \text{ and } K\gamma^*(\pi)K = \cup_{j=1}^{J} y_j K,$$

where both I and J are finite (since $q < \infty$). The theorem will be a consequence of the following lemma.

LEMMA 8.10. $m_{\alpha,\beta,\gamma}(0) = \#(\mathcal{T}_{\alpha,\beta,\gamma})$.

Proof: We have

$$\begin{aligned}
m_{\alpha,\beta,\gamma}(0) = c_\alpha \cdot c_\beta \cdot c_\gamma(1) &= \int_G (\int_G c_\alpha(x) c_\beta(x^{-1}y) c_\gamma(y^{-1}) dx) dy \\
&= \int_G \left(\sum_{i=1}^{I} \int_K c_\alpha(x_i k) c_\beta(k^{-1} x_i^{-1} y) dk \right) c_\gamma(y^{-1}) dy \\
&= \sum_{i=1}^{I} c_\alpha(x_i) \int_G c_\beta(x_i^{-1} y) c_\gamma(y^{-1}) dy \\
&= \sum_{i=1}^{I} \sum_{j=1}^{J} c_\alpha(x_i) \int_K c_\beta(x_i^{-1} y_j k) c_\gamma(k^{-1} y_j^{-1}) dk \\
&= \sum_{i=1}^{I} \sum_{j=1}^{J} c_\alpha(x_i) c_\beta(x_i^{-1} y_j) c_\gamma(y_j^{-1}) \\
&= \sum_{i=1}^{I} \sum_{j=1}^{J} c_\beta(x_i^{-1} y_j) = \#(\mathcal{T}_{\alpha,\beta,\gamma}) \quad \square
\end{aligned}$$

As a consequence of Theorem 8.1 and Lemma 8.5, we find that the structure constants for the Hecke algebra are determined by the geometry of the building.

THEOREM 8.2. *Let $\alpha, \beta, \gamma \in X_*(\underline{T})$. Then we have*

$$m_{\alpha,\beta}(\gamma) = \frac{\#(\mathcal{T}_{\alpha,\beta,\gamma^*})}{\#(S(o,\gamma(\pi)))}. \tag{8.2}$$

Proof: We have only to apply Lemma 8.5 and observe that since we are assuming that $vol(K) = 1$ we have $vol(K\gamma(\pi)K) = vol(K\gamma(\pi)K/K) = \#(S(o,\gamma(\pi)))$. □

8.3. The Satake transform

In this section we define an integral transform S, the Satake transform, from compactly supported, K-biinvariant functions on G to left K-invariant, right U-invariant functions on G.

Let $\delta : B \longrightarrow \mathbb{R}_+^*$ be the modular function of B, [**Gro**, §3]. We may regard δ as a left K-invariant, right U-invariant function on G. By the Iwasawa decomposition for G, [**Tits**, pg. 51], any such function is determined by its restriction to T. We normalize the Haar measure du on U so that the open subgroup $K \cap U$ has measure 1. For a compactly supported K-biinvariant function f on G we define its *Satake transform* as a function $Sf(g)$ on G given by

$$Sf(g) = \delta(g)^{1/2} \cdot \int_U f(gu)du.$$

Then Sf is a left K-invariant, right U-invariant function on G with values in $\mathbb{Z}[q^{1/2}, q^{-1/2}]$; this function is determined by its restriction to $T/T \cap K \cong X_*(\underline{T})$. The main result of [**Sat**] (see also [**C**, pg. 147]), is that the image of S lies in the subring

$$(\mathbb{Z}[X_*(\underline{T})])^W \otimes \mathbb{Z}[q^{1/2}, q^{-1/2}] \cong R(G^\vee) \otimes \mathbb{Z}[q^{1/2}, q^{-1/2}],$$

where $\cong$ is a ring isomorphism. Here and below, $G^\vee = \underline{G}^\vee(\mathbb{C})$. Furthermore we have (see [**Sat**], [**C**, pg. 147]):

THEOREM 8.3. *The Satake transform gives a ring isomorphism*

$$S : \mathcal{H} \otimes \mathbb{Z}[q^{1/2}, q^{-1/2}] \cong R(G^\vee) \otimes \mathbb{Z}[q^{1/2}, q^{-1/2}].$$

For $\lambda \in P^+$ let chV_λ be the character of the irreducible representation V_λ of $G^\vee$ (see e.g. [**FH**, pg. 375]). We may identify chV_λ with a W-invariant element of $\mathbb{Z}[X^*(\underline{T}^\vee)] = \mathbb{Z}[X_*(\underline{T})]$.

In what follows we will need two bases for the free $\mathbb{Z}[q^{1/2}, q^{-1/2}]$-module

$$R(G^\vee) \otimes \mathbb{Z}[q^{1/2}, q^{-1/2}].$$

The first basis is

$$\mathcal{S} := \{S(c_\lambda) : \lambda \in P^+\},$$

the second basis is

$$\mathcal{R} := \{chV_\lambda : \lambda \in P^+\}.$$

The change of basis matrices relating these two bases are both upper triangular (see Lemma 8.11 below) with entries in the ring $\mathbb{Z}[q^{1/2}, q^{-1/2}]$. We define the order (at ∞) $ord(f)$ of the element

$$f = \sum_{k=-M}^{k=N} a_k q^{k/2} \in \mathbb{Z}[q^{1/2}, q^{-1/2}]$$

by $ord(f) = N$, provided that $a_N \neq 0$. Note that if f is a polynomial in q then $ord(f) = 2deg(f)$ where $deg(f)$ is the degree of f in q. We will accordingly extend the degree to $\mathbb{Z}[q^{1/2}, q^{-1/2}]$ by defining

$$deg(f) = 1/2\ ord(f).$$

Thus the extended degree takes values in the half integers. We define $deg_{\mathcal{S}}(F)$ for

$$F \in R(G^\vee) \otimes \mathbb{Z}[q^{1/2}, q^{-1/2}]$$

to be the maximum of the degrees of the components of F *when F is expressed in the basis $\mathcal{S}$*. We will use a similar convention (i.e., expanding it terms of the basis $\mathcal{S}$) when we speak of the "leading term" of F. We retain the notation deg (without subscript) for the ordinary notion of degree $\in \frac{1}{2}\mathbb{Z}$ for the Laurent polynomials in $\mathbb{Z}[q^{1/2}, q^{-1/2}]$. Analogously, we define $deg_{\mathcal{R}}(F)$ when F is expanded in terms of the basis $\mathcal{R}$.

In what follows we will need two formal properties of $deg_{\mathcal{R}}$ and $deg_{\mathcal{S}}$. We leave their proofs to the reader. The first is the *ultrametric* inequality which we state for $deg_{\mathcal{S}}$

$$deg_{\mathcal{S}}(\sum_{i=1}^{n} F_i) \leq \max_{1\leq i\leq n}(deg_{\mathcal{S}}(F_1), \cdots, deg_{\mathcal{S}}(F_n)).$$

The second is the obvious identity (again stated for $deg_{\mathcal{S}}$)

$$deg_{\mathcal{S}}(p \cdot F) = deg(p) + deg_{\mathcal{S}}(F),\ p \in \mathbb{Z}[q^{1/2}, q^{-1/2}].$$

For the following lemma, due to Lusztig, we refer the reader to [**Gro**, (3.11) and (3.12)], [**Lu2**, (6.10)]. See [**Ha1**, §2], for the statement of the lemma in the generality we require here.

LEMMA 8.11. *1. There exist polynomials $a_\lambda(\mu)$ in q such that*

$$S(c_\lambda) = q^{\langle\lambda,\rho\rangle} chV_\lambda + \sum_{\mu<\lambda} a_\lambda(\mu) q^{\langle\mu,\rho\rangle} chV_\mu.$$

2. Conversely, there exist polynomials $b_\lambda(\mu)$ in q such that

$$q^{\langle\lambda,\rho\rangle} chV_\lambda = S(c_\lambda) + \sum_{\mu<\lambda} b_\lambda(\mu) S(c_\mu).$$

The degree estimate below follows from [**Lu2**], who proved that the polynomials $b_\lambda(\mu)$ are *Kazhdan-Lusztig polynomials.* These polynomials have many remarkable properties but the only property we need here is the degree estimate. The inequality of the next lemma will play a critical role in our proofs.

LEMMA 8.12. *For all $\mu < \lambda, \mu \in P^+$ we have:*

$$deg(b_\lambda(\mu)) < \langle\lambda - \mu, \rho\rangle \leq \langle\lambda, \rho\rangle.$$

This lemma when combined with Lemma 8.11 has the following consequences.

LEMMA 8.13.

(1) $deg_{\mathcal{S}}(chV_\lambda) < 0$, *for all $\lambda \in P^+ \setminus \{0\}$* .
(2) $deg_{\mathcal{S}} \leq deg_{\mathcal{R}}$.

Proof: By Lemma 8.11 we get

$$chV_\lambda = q^{-\langle\lambda,\rho\rangle}S(c_\lambda) + \sum_{\mu<\lambda} q^{-\langle\lambda,\rho\rangle}b_\lambda(\mu)S(c_\mu).$$

Hence Lemma 8.12 implies

$$deg_{\mathcal{S}}(chV_\lambda) < 0.$$

Since $deg_{\mathcal{R}}(chV_\lambda) = 0$ we find that the inequality of the second statement of the lemma holds on a basis. Consequently it holds for all

$$F \in \mathcal{H} \otimes \mathbb{Z}[q^{1/2}, q^{-1/2}] \cong R(G^\vee) \otimes \mathbb{Z}[q^{1/2}, q^{-1/2}].$$

Indeed, writing $F = \sum_{i=1}^n p_i \; chV_i$ we obtain, using the ultrametric inequality and the first statement of the lemma,

$$\begin{aligned} deg_{\mathcal{S}}(F) &\leq \max_{1\leq i\leq n} (deg_{\mathcal{S}}(p_1 chV_1), \cdots, deg_{\mathcal{S}}(p_n chV_n)) \\ &\leq (deg(p_1), \cdots, deg(p_n)) = deg_{\mathcal{R}}(F). \end{aligned}$$

□

Recall that 0 denote the trivial cocharacter.

LEMMA 8.14. *There exists a polynomial $M_{\alpha,\beta,\gamma}(q)$ in the variable q, such that we have the equality of functions in q:*

$$m_{\alpha,\beta,\gamma}(0) = M_{\alpha,\beta,\gamma}(q).$$

Furthermore, all structure constants $m_{\alpha,\beta}(\gamma)$ are polynomials in q.

Proof: We will prove that all the structure constants $m_{\alpha,\beta}(\gamma)$ are polynomials in q. The first statement will follow from this.

Following [**Gro**] we define $\phi_\lambda \in R(G^\vee) \otimes \mathbb{Z}[q^{1/2}, q^{-1/2}]$ by

$$\phi_\lambda = q^{\langle\lambda,\rho\rangle} chV_\lambda.$$

The elements $\{\phi_\lambda : \lambda \in X^*(\underline{T}^\vee)\}$ give a new basis for the $\mathbb{Z}[q^{1/2}, q^{-1/2}]$-module $R(G^\vee) \otimes \mathbb{Z}[q^{1/2}, q^{-1/2}]$.

We claim that the structure constants for the ring relative to this basis are polynomials in q. Indeed, we have

$$\begin{aligned} \phi_\alpha \cdot \phi_\beta &= q^{\langle\alpha+\beta,\rho\rangle} chV_\alpha \cdot chV_\beta = q^{\langle\alpha+\beta,\rho\rangle} \sum_{\gamma\leq\alpha+\beta} n_{\alpha+\beta}(\gamma) \; chV_\gamma \\ &= \sum_{\gamma\leq\alpha+\beta} n_{\alpha+\beta}(\gamma) \; q^{\langle\alpha+\beta-\gamma,\rho\rangle} \phi_\gamma. \end{aligned}$$

Now, since $\gamma \leq \alpha+\beta$, the coweight $\alpha+\beta-\gamma$ is a sum of positive coroots and consequently $\langle\alpha+\beta-\gamma,\rho\rangle$ is an nonnegative integer.

We can now prove the second statement. Let α and β be given. We expand $S(c_\alpha)$ and $S(c_\beta)$ in terms of the basis of ϕ_λ's thereby introducing the polynomials $a_\lambda(\mu)$. We then multiply the resulting expressions. According to the paragraph above, the result is an expression in the ϕ_λ's with polynomial coefficients in q. We then substitute for the ϕ_λ's using Lemma 8.11 introducing the polynomials $b_\lambda(\mu)$. □

8.3.1. A remarkable nonvanishing property of the polynomials $M_{\alpha,\beta,\gamma}$. It follows from Theorem 8.1 that if q is a prime power then the values $M_{\alpha,\beta,\gamma}(q)$ are nonnegative. We now show that if for some prime power q the value $M_{\alpha,\beta,\gamma}(q)$ is nonzero then all values at prime powers are nonzero. We note that the examples we computed below show that the polynomial $M_{\alpha,\beta,\gamma}$ can have coefficients of both signs.

THEOREM 8.4. *If a polynomial $M_{\alpha,\beta,\gamma}$ is nonzero for some prime power $q = p^e$ then it is nonzero at all prime powers.*

Proof: We assume that q is given such that

$$M_{\alpha,\beta,\gamma}(q) \neq 0.$$

This means that

$$(\alpha, \beta, \gamma) \in Sol(\mathbf{Q3}, \underline{G}(\mathbb{K}_q)),$$

where $\mathbb{K}_q$ is a local field with residue field of order q. In other words, in the Euclidean building X_q corresponding to the group $\underline{G}(\mathbb{K}_q)$ there exists a triangle τ with Δ-side lengths α, β, γ, whose vertices are special vertices of X. It then follows from Part 5 of Theorem 1.2 or, equivalently, from the Transfer Theorem 4.1, that

$$(\alpha, \beta, \gamma) \in Sol(\mathbf{Q3}, \underline{G}(\mathbb{K})),$$

for an *arbitrary* local field $\mathbb{K}$ (with any order $q' \geq 2$ of the residue field). Therefore $M_{\alpha,\beta,\gamma}(q') \neq 0$ for all prime powers q'. □

8.4. A solution of Problem Q4 is a solution of Problem Q3

We recall the definition of the structure constants $n_{\alpha,\beta,\gamma}(\delta)$:

$$ch(V_\alpha) \cdot ch(V_\beta) \cdot ch(V_\gamma) = \sum_{\delta \in P^+} n_{\alpha,\beta,\gamma}(\delta)\; ch(V_\delta).$$

We are interested in comparing the coefficient $n_{\alpha,\beta,\gamma}(0)$ (corresponding to the trivial character 0 of $G^\vee$) with $m_{\alpha,\beta,\gamma}(0)$.

The main goal of this section is to prove the following

THEOREM 8.5. *If the triple α, β, γ is a solution of Problem* **Q4** *then it is also a solution of Problem* **Q3**. *More precisely,*

$$n_{\alpha,\beta,\gamma}(0) \neq 0 \Longrightarrow m_{\alpha,\beta,\gamma}(0) \neq 0.$$

We first prove the following:

THEOREM 8.6. *Suppose that $\underline{G}$ is of a reductive algebraic group over $\mathbb{K}$ which is split over $\mathbb{K}$. Then:*

(a) The degree of the polynomial $M_{\alpha,\beta,\gamma}(q)$ is at most $\langle \alpha + \beta + \gamma, \rho \rangle$.

(b) The leading coefficient of $M_{\alpha,\beta,\gamma}(q)$, i.e., the coefficient at $q^{\langle \alpha+\beta+\gamma, \rho\rangle}$, is equal to $n_{\alpha,\beta,\gamma}(0)$.

Proof of Theorem 8.6. For the benefit of the reader, below we explain the idea behind the proof of Theorem 8.6. We have to compute the coefficient of 0 in the Hecke triple product $c_\alpha \cdot c_\beta \cdot c_\gamma$. Instead, we will compute the coefficient of 0 in the triple product $S(c_\alpha) \cdot S(c_\beta) \cdot S(c_\gamma)$ in the representation ring $R(G^\vee) \otimes \mathbb{Z}[q^{1/2}, q^{-1/2}]$ where the triple product is expanded *relative to the basis $\mathcal{S}$*. In order to do this we will use the formula from Lemma 8.11 and compute (initially) in the basis $\mathcal{R}$.

However, to prove the theorem we must again apply Lemma 8.11 to rewrite the result in terms of the basis $\mathcal{S}$.

The theorem will follow from the next three lemmas. We owe the first of these to Jiu-Kang Yu.

LEMMA 8.15. *For all $\mu \in P^+$ satisfying $\mu < \lambda$ we have:*

$$deg(a_\lambda(\mu)q^{\langle \mu,\rho\rangle}) < \langle \lambda, \rho\rangle.$$

Proof: The proof is by induction on λ with respect to the partial order $<$. We remind the reader that each set $\{\mu \in P^+ : \mu < \lambda\}$ is finite. If the set $\{\mu \in P^+ : \mu < \lambda\}$ is empty then there is nothing to prove.

Now assume that we are given $\lambda \in P^+$ and that we have proved the above estimate for all predecessors of λ in the partial order $<$. We have, according to Lemma 8.11,

$$S(c_\lambda) = q^{\langle \lambda,\rho\rangle} chV_\lambda - \sum_{\mu<\lambda} b_\lambda(\mu) S(c_\mu)$$

and

$$S(c_\mu) = q^{\langle \mu,\rho\rangle} chV_\mu + \sum_{\eta<\mu} a_\mu(\eta) q^{\langle \eta,\rho\rangle} chV_\eta.$$

Thus

$$S(c_\lambda) = q^{\langle \lambda,\rho\rangle} chV_\lambda - \sum_{\mu<\lambda} b_\lambda(\mu) q^{\langle \mu,\rho\rangle} chV_\mu - \sum_{\mu<\lambda}\sum_{\eta<\mu} b_\lambda(\mu) a_\mu(\eta) q^{\langle \eta,\rho\rangle} chV_\eta.$$

We make the change of variable $\mu \to \tau, \eta \to \mu$ in the last sum to obtain

$$S(c_\lambda) = q^{\langle \lambda,\rho\rangle} chV_\lambda - \sum_{\mu<\lambda} b_\lambda(\mu) q^{\langle \mu,\rho\rangle} chV_\mu - \sum_{\mu<\lambda}\ \sum_{\{\tau:\mu<\tau<\lambda\}} b_\lambda(\tau) a_\tau(\mu) q^{\langle \mu,\rho\rangle} chV_\mu.$$

Since $chV_\mu, \mu \in P^+$, is a basis of the $\mathbb{Z}[q^{-1/2}, q^{1/2}]$-module

$$R(G^\vee) \otimes \mathbb{Z}[q^{1/2}, q^{-1/2}],$$

by combining the previous formula with the part 1 of Lemma 8.11, we obtain:

$$a_\lambda(\mu) q^{\langle \mu,\rho\rangle} = -b_\lambda(\mu) q^{\langle \mu,\rho\rangle} - \sum_{\{\tau:\mu<\tau<\lambda\}} b_\lambda(\tau) a_\tau(\mu) q^{\langle \mu,\rho\rangle}.$$

The degree of the first term satisfies the required estimate by Lemma 8.12. We estimate the degrees of the terms in the second sum. By the induction hypothesis, $deg(a_\tau(\mu)q^{\langle \mu,\rho\rangle}) < \langle \tau, \rho\rangle$ and by Lemma 8.12 we have

$$deg(b_\lambda(\tau)) < \langle \lambda - \tau, \rho\rangle.$$

Consequently, for all μ, we get

$$deg(b_\lambda(\tau) a_\tau(\mu) q^{\langle \mu,\rho\rangle}) < \langle \lambda - \tau, \rho\rangle + \langle \tau, \rho\rangle = \langle \lambda, \rho\rangle. \quad \square$$

COROLLARY 8.16. *$deg(a_\lambda(\mu)) < \langle \lambda - \mu, \rho\rangle$.*

LEMMA 8.17. *For $\alpha, \beta, \gamma \in P^+$ such that $\alpha + \beta + \gamma$ is a nontrivial character, we have:*

(1) $deg_{\mathcal{R}}(S(c_\alpha) \cdot S(c_\beta) \cdot S(c_\gamma) - q^{\langle \alpha+\beta+\gamma,\rho\rangle} chV_\alpha \cdot chV_\beta \cdot chV_\gamma) < \langle \alpha + \beta + \gamma, \rho\rangle$.
(2) $deg_{\mathcal{R}}(S(c_\alpha) \cdot S(c_\beta) \cdot S(c_\gamma)) = \langle \alpha + \beta + \gamma, \rho\rangle$.

Proof: By expanding the triple product

$$S(c_\alpha) \cdot S(c_\beta) \cdot S(c_\gamma) =$$

$$(q^{\langle\gamma,\rho\rangle} chV_\gamma + \sum_{\lambda<\gamma} a_\gamma(\lambda) q^{\langle\lambda,\rho\rangle} chV_\lambda)$$
$$\cdot(q^{\langle\beta,\rho\rangle} chV_\beta + \sum_{\mu<\beta} a_\beta(\mu) q^{\langle\mu,\rho\rangle} chV_\mu)$$
$$\cdot(q^{\langle\alpha,\rho\rangle} chV_\alpha + \sum_{\nu<\alpha} a_\alpha(\nu) q^{\langle\nu,\rho\rangle} chV_\nu)$$

in terms of the basis $\mathcal{R} := \{chV_\tau\}$ we get the following types of summands:

$$q^{\langle\alpha+\beta+\gamma,\rho\rangle} chV_\alpha \cdot chV_\beta \cdot chV_\gamma,$$

$$q^{\langle\alpha+\beta,\rho\rangle} chV_\alpha \cdot chV_\beta \sum_{\lambda<\gamma} a_\gamma(\lambda) q^{\langle\lambda,\rho\rangle} chV_\lambda,$$

$$q^{\langle\gamma,\rho\rangle} chV_\gamma \cdot (\sum_{\mu<\beta} a_\beta(\mu) q^{\langle\mu,\rho\rangle} chV_\mu) \cdot (\sum_{\nu<\alpha} a_\alpha(\nu) q^{\langle\nu,\rho\rangle} chV_\nu),$$

and similar ones obtained by permuting α, β, γ, and finally:

$$a_\alpha(\nu) a_\beta(\mu) a_\gamma(\lambda) q^{\langle\nu+\mu+\lambda,\rho\rangle} chV_\alpha \cdot chV_\beta \cdot chV_\gamma,$$

where in the latter case $\nu < \alpha, \mu < \beta, \lambda < \gamma$, $\nu, \mu, \lambda \in P^+$. The first term

$$q^{\langle\alpha+\beta+\gamma,\rho\rangle} chV_\alpha \cdot chV_\beta \cdot chV_\gamma$$

cancels out, we estimate the degree of each of the remaining terms separately. We will do it in the case of the 2-nd and 4-th term and leave the 3-rd term to the reader:

First, since $\lambda < \gamma$,

$$\begin{aligned} deg_{\mathcal{R}}[q^{\langle\alpha+\beta,\rho\rangle} chV_\alpha \cdot chV_\beta \cdot a_\gamma(\lambda) q^{\langle\lambda,\rho\rangle} chV_\lambda] \leq \\ \langle\alpha+\beta,\rho\rangle + deg(a_\gamma(\lambda) q^{\langle\lambda,\rho\rangle}) < \\ \text{(by Lemma 8.15)} \\ < \langle\alpha+\beta,\rho\rangle + \langle\gamma,\rho\rangle = \langle\alpha+\beta+\gamma,\rho\rangle. \end{aligned}$$

Similarly,

$$\begin{aligned} deg_{\mathcal{R}}[a_\alpha(\nu) a_\beta(\mu) a_\gamma(\lambda) q^{\langle\nu+\mu+\lambda,\rho\rangle} chV_\alpha \cdot chV_\beta \cdot chV_\gamma] = \\ deg[a_\alpha(\nu) q^{\langle\nu,\rho\rangle}] + deg[a_\beta(\mu) q^{\langle\mu,\rho\rangle}] + deg[a_\gamma(\lambda) q^{\langle\lambda,\rho\rangle}] < \\ \text{(by Lemma 8.15)} \\ < \langle\alpha+\beta+\gamma,\rho\rangle. \end{aligned}$$

The first statement immediately implies the second and the lemma follows. □

COROLLARY 8.18. *Under the assumptions of the above lemma, we have:*

$$deg_{\mathcal{S}}(S(c_\alpha) \cdot S(c_\beta) \cdot S(c_\gamma) - q^{\langle\alpha+\beta+\gamma,\rho\rangle} chV_\alpha \cdot chV_\beta \cdot chV_\gamma) < \langle\alpha+\beta+\gamma,\rho\rangle.$$

Proof: The assertion follows from Lemma 8.17 since by Lemma 8.13 we have

$$deg_{\mathcal{S}} \leq deg_{\mathcal{R}}.$$

□

Now we can prove the degree estimate in Theorem 8.6. Observe that since $M_{\alpha,\beta,\gamma}(q)$ is one of the coefficients (the coefficient of the trivial double coset) of the triple product $c_\alpha \cdot c_\beta \cdot c_\gamma$ when expressed in terms of the double coset basis for the Hecke ring, we have

$$deg(M_{\alpha,\beta,\gamma}(q)) \leq deg_{\mathcal{S}}(c_\alpha \cdot c_\beta \cdot c_\gamma)$$

and hence

$$\begin{aligned} deg(M_{\alpha,\beta,\gamma}(q)) &\leq deg_{\mathcal{S}}(S(c_\alpha) \cdot S(c_\beta) \cdot S(c_\gamma)) \leq \\ & deg_{\mathcal{R}}(S(c_\alpha) \cdot S(c_\beta) \cdot S(c_\gamma)) = \langle \alpha + \beta + \gamma, \rho \rangle. \end{aligned}$$

Here the last equality follows from Lemma 8.17. This proves the first assertion of Theorem 8.6.

We now identify the leading term of $M_{\alpha,\beta,\gamma}(q)$.

Let $N_{\alpha,\beta,\gamma}$ be the element of the Hecke ring defined by

$$N_{\alpha,\beta,\gamma} := q^{\langle \alpha+\beta+\gamma,\rho \rangle} chV_\alpha \cdot chV_\beta \cdot chV_\gamma,$$

Let $\mathbb{I}$ denote the identity element of the representation ring (i.e. the character of the trivial representation). We identify this element (via the Satake isomorphism) with the trivial double coset c_0. We have

LEMMA 8.19.

$$deg_{\mathcal{S}}(N_{\alpha,\beta,\gamma} - n_{\alpha,\beta,\gamma}(0) q^{\langle \alpha+\beta+\gamma,\rho \rangle} \mathbb{I}) < \langle \alpha + \beta + \gamma, \rho \rangle.$$

Proof: By definition of the structure constants $n_{\alpha,\beta,\gamma}(\delta)$ we have

$$q^{\langle \alpha+\beta+\gamma,\rho \rangle} chV_\alpha \cdot chV_\beta \cdot chV_\gamma = \sum_{\delta \leq \alpha+\beta+\gamma} n_{\alpha,\beta,\gamma}(\delta)\, q^{\langle \alpha+\beta+\gamma-\delta,\rho \rangle} q^{\langle \delta,\rho \rangle} chV_\delta.$$

But by Lemma 8.13,

$$\delta \neq 0 \Longrightarrow deg_{\mathcal{S}}(q^{\langle \delta,\rho \rangle} chV_\delta) < \langle \delta, \rho \rangle.$$

Noting that $\mathbb{I}$ is the character of the trivial representation the lemma follows. □

We now combine the previous lemma, the ultrametric inequality and Corollary 8.18 to obtain the estimate

$$\begin{aligned} & deg_{\mathcal{S}}(S(c_\alpha) \cdot S(c_\beta) \cdot S(c_\gamma)) - n_{\alpha,\beta,\gamma}(0) q^{\langle \alpha+\beta+\gamma,\rho \rangle} \mathbb{I}) \\ & \leq \max(deg_{\mathcal{S}}(S(c_\alpha) \cdot S(c_\beta) \cdot S(c_\gamma) - N_{\alpha,\beta,\gamma}),\ deg_{\mathcal{S}}(N_{\alpha,\beta,\gamma} - n_{\alpha,\beta,\gamma}(0) q^{\langle \alpha+\beta+\gamma,\rho \rangle} \mathbb{I})) \\ & < \langle \alpha + \beta + \gamma, \rho \rangle. \end{aligned}$$

Now the previous estimate on the difference of elements in the Hecke ring implies that degrees of *all* coefficient polynomials of the difference when expanded in terms of the double coset basis satisfy the same estimate. Now the coefficient of the identity element $c_0 = \mathbb{I}$ of the triple product is $m_{\alpha,\beta,\gamma}(0) = M_{\alpha,\beta,\gamma}(q)$. Thus applying the above observation to the identity coefficient of the difference we obtain

$$deg(M_{\alpha,\beta,\gamma}(q) - n_{\alpha,\beta,\gamma}(0) q^{\langle \alpha+\beta+\gamma,\rho \rangle}) < \langle \alpha + \beta + \gamma, \rho \rangle.$$

This proves Theorem 8.6 and in particular implies that the polynomial $M_{\alpha,\beta,\gamma}$ is nonzero. Hence for nonarchimedean local fields with sufficiently large residue fields (i.e. sufficiently large q) the structure constant $m_{\alpha,\beta,\gamma}(0)$ is nonzero. But by Theorem 8.4 if the structure constant $m_{\alpha,\beta,\gamma}(0)$ is nonzero for some nonarchimedean local field $\mathbb{K}$ it is nonzero for all nonarchimedean local fields. Hence Theorem 8.5 is proved.

The above theorem also immediately implies the following

COROLLARY 8.20. *If* $n_{\alpha,\beta,\gamma}(0) \neq 0$ *then* $\langle \alpha + \beta + \gamma, \rho \rangle \in \mathbb{Z}$.

We now give another explanation why

$$n_{\alpha,\beta,\gamma}(0) \neq 0 \Longrightarrow \langle \alpha + \beta + \gamma, \rho \rangle \in \mathbb{Z}.$$

THEOREM 8.7.

(1) $$m_{\alpha,\beta,\gamma}(0) \neq 0 \Longrightarrow \alpha + \beta + \gamma \in Q(R^\vee) \Longrightarrow \langle \alpha + \beta + \gamma, \rho \rangle \in \mathbb{Z}.$$

(2) $$n_{\alpha,\beta,\gamma}(0) \neq 0 \Longrightarrow \alpha + \beta + \gamma \in Q(R^\vee).$$

Proof: (1) follows from Theorem 7.1, part 4. Hence (2) follows from (1) via Theorem 8.6, however we give a direct proof below.

The center of $G^\vee$ acts on each of the factors of the triple tensor product by a scalar, hence it acts by a scalar on the triple tensor product itself. But if $n_{\alpha,\beta,\gamma}(0) \neq 0$ this scalar is necessarily 1. Hence the center fixes the (highest) weight vector with weight $\alpha + \beta + \gamma$, whence $\alpha + \beta + \gamma$ annihilates the center. Accordingly it is in the character lattice of the adjoint group $Ad\ G^\vee$. Thus $\alpha + \beta + \gamma \in Q(R^\vee)$. □

8.5. A Solution of Problem Q3 is not necessarily a solution of Problem Q4

In this section we consider two examples: $\underline{G} = SO(5)$, whence $G^\vee = Sp(4, \mathbb{C})$ and the group $\underline{G}$ of type G_2 and $G^\vee = G_2(\mathbb{C})$.

We begin with $\underline{G} = SO(5)$; we assume that we have chosen Witt bases for $\mathbb{C}^5$ and $\mathbb{C}^4$. We let $T^\vee$ be the split torus consisting of diagonal matrices in $Sp(4, \mathbb{C})$ and $\underline{T}$ be the split torus of $SO(5)$ consisting of diagonal matrices. We use the rectangular coordinates x_1, x_2 in the Cartan subalgebra $\mathfrak{a}^\vee \subset \mathfrak{sp}(4, \mathbb{C})$ such that the simple roots are $x_1 - x_2, 2x_2$. We set

$$\alpha = \beta = \gamma = (1, 1) \in X^*(\underline{T}^\vee) = X_*(\underline{T}).$$

Note that $\alpha+\beta+\gamma$ belongs to the coroot lattice of $SO(5)$, i.e., the condition stated in Theorem 8.7 is satisfied.

We give two proofs of the next lemma, the first computational, the second conceptual.

LEMMA 8.21.

$$n_{\alpha,\beta,\gamma}(0) = 0.$$
$$m_{\alpha,\beta,\gamma}(0) = q^5 - q.$$

Thus this choice of side-lengths is a solution Problem **Q3** *but not of Problem* **Q4**.

REMARK 8.22. In fact the second equality implies the first because, in this case, $\langle \alpha+\beta+\gamma, \rho\rangle = 6$. Hence, (according to Theorem 8.6)

$$M_{\alpha,\beta,\gamma}(q) = n_{\alpha,\beta,\gamma}(0)q^6 + \text{lower terms}.$$

But the coefficient of q^6 in the above formula for $m_{\alpha,\beta,\gamma}(0)$ is 0.

Proof: The first equality is obvious if we observe that under the isomorphism between $Sp(4)$ and $Spin(5)$ the representation of $Sp(4,\mathbb{C})$ with the highest weight $(1,1)$ corresponds to the standard (vector) representation of $SO(5,\mathbb{C})$ on $\mathbb{C}^5$ (considered as a nonfaithful representation of $Spin(5,\mathbb{C})$). Now it is standard that the tensor square of this representation consists of three irreducible summands, the trivial representation (corresponding to the invariant quadratic form), the exterior square of the standard representation and the harmonic quadratic polynomials on $\mathbb{C}^5$. Since none of these are equivalent to the standard representation and all representations of $Sp(4,\mathbb{C})$ are self-dual the first statement follows.

The proof of the second statement requires more work. From [**Gro**, pg. 231], we have:

$$S(c_{(1,1)}) = q^2 chV_{(1,1)} - 1.$$

Upon taking the cube and calculating in the representation ring of $Sp(4,\mathbb{C})$ one obtains

$$\begin{aligned} S(c_{(1,1)})^3 &= q^6[chV_{(3,3)} + 2chV_{(3,1)} + chV_{(2,0)} + 3chV_{(1,1)}] \\ &\quad -3q^4[chV_{(2,2)} + chV_{(2,0)} + 1] + 3q^2 chV_{(1,1)} - 1. \end{aligned}$$

Upon substituting for the chV_λ using Lemma 8.11 (and computing the appropriate Kazhdan-Lusztig polynomials $b_\lambda(\mu)$) the lemma follows. □

As for the conceptual proof of (the second part of) the lemma, note that the side-lengths belong to the root-subgroup $SL(2)$ corresponding to the positive root $x_1 + x_2$. But it is evident that one can construct an equilateral triangle with side-lengths equal to the positive root and vertices of the correct type in the tree for $SL(2,\mathbb{K})$. Since this tree is convex in the building for $SO(5,\mathbb{K})$ and the fixed vertex for $SL(2,\mathcal{O})$ in this embedded tree is o (the fixed vertex for $SO(5,\mathcal{O})$), the lemma follows. Equivalently if we can realize the trivial double coset as a product of the three $SL(2,\mathcal{O})$–double cosets in $SL(2,\mathbb{K})$ belonging to $(1,1),(1,1),(1,1)$ then we can certainly do it in $SO(5,\mathbb{K})$.

REMARK 8.23. This last sentence is the essence of the counterexample. The solutions to Problems **Q3** and **Q4** behave differently with respect to inclusions of subgroups. The solutions of Problem **Q3** "push forward", the solutions of Problem **Q4** do not.

We now give another example of a triple of coroots α,β,γ such that the triple is a solution to Problem **Q3** but not of Problem **Q4**. This example was motivated by unpublished observations of S. Kumar and J. Stembridge.

We take $\underline{G} = G_2$. Hence $G^\vee = G_2$ as well. We take $\alpha = \lambda_1$ and $\beta = \gamma = \lambda_2$. Here λ_1 is the first fundamental weight (the highest weight of the unique irreducible seven dimensional representation) and λ_2 is the second fundamental weight (the highest weight of the adjoint representation).

LEMMA 8.24. $m_{\lambda_1,\lambda_2,\lambda_2}(0) = q^5(q+1)(q^6-1)$.

Proof: From [**Gro**, pg. 231, (5.7)], we have

$$\begin{aligned} S(c_{\lambda_1}) &= q^3 chV_{\lambda_1} - 1 \\ S(c_{\lambda_2}) &= q^5 chV_{\lambda_2} - q^3 chV_{\lambda_2} - q^4. \end{aligned}$$

Multiplying (using LiE) one obtains

$$S(c_{\lambda_1})\cdot S(c_{\lambda_2}) = q^8 chV_{\lambda_1+\lambda_2} + (q^8-q^6)(chV_{2\lambda_1} + chV_{\lambda_1}) - q^6 chV_{\lambda_2} - q^4 S(c_{\lambda_1}) - S(c_{\lambda_2}).$$

Hence, by Lemma 8.11 we have

$$m_{\lambda_1,\lambda_2}(\lambda_2) = b_{\lambda_2,\lambda_1+\lambda_2}(q) + (q^2-1)b_{\lambda_2,2\lambda_1}(q) - q - 1.$$

Using $b_{\lambda_2,\lambda_1+\lambda_2}(q) = 1+q$ and $b_{\lambda_2,2\lambda_1}(q) = 1$ we obtain

$$m_{\lambda_1,\lambda_2}(\lambda_2) = q^2 - 1.$$

Now Lemma 8.5 implies that

$$m_{\lambda_1,\lambda_2,\lambda_2}(0) = m_{\lambda_1,\lambda_2}(\lambda_2)\cdot vol(K\lambda_2(\pi)K).$$

From [**Gro**, pg. 735], we get $vol(K\lambda_2(\pi)K) = deg(c_{\lambda_2}) = q^{10}+q^9+q^8+q^7+q^6+q^5$. The lemma follows. □

COROLLARY 8.25.

$$n_{\lambda_1,\lambda_2,\lambda_2}(0) = 0.$$

Proof: From [**Gro**, pg. 231, (5.7)], we have $\langle\lambda_1,\rho\rangle = 3$ and $\langle\lambda_2,\rho\rangle = 5$ whence $\langle\lambda_1+\lambda_2+\lambda_2,\rho\rangle = 13$. Hence, the structure constant has degree (one) less than the maximum possible degree. □

We now give a geometric proof (using unfolding) that $m_{\lambda_1,\lambda_2,\lambda_2}(0)$ is nonzero. We show that there exists a triangle T' (with Δ-side-lengths $\lambda_1,\lambda_2,\lambda_2$) in a discrete Euclidean building X modeled on the discrete affine Coxeter complex of type G_2, so that all vertices of T' are special. We start with the billiard triangle T in Figure 1, contained in a model apartment $A \subset X$; this triangle has two geodesic sides $\overline{zx}$, $\overline{zy}$ and one broken side $\overline{xuy}$. Note that all three vertices of this billiard triangle are special vertices of the Coxeter complex. The break point u on the broken side of the billiard triangle belongs to the wall l of the Coxeter complex, the wall l separates the broken side $\overline{xuy}$ from the vertex $z = o$. Thus, by Lemma 6.7, one can unfold the billiard triangle T to a geodesic triangle $T' \subset X$ preserving the refined side-lengths. One can also easily see that $n_{\lambda_1,\lambda_2,\lambda_2}(0) = 0$ using Littelmann's path model.

8.6. The saturation theorem for $GL(\ell)$

In this subsection we explain Green's idea for proving that for $GL(\ell)$

$$Sol(\mathbf{Q3}) \subset Sol(\mathbf{Q4}).$$

Let $\mathbb{K}$ be a valued field with a discrete complete valuation v and let $\mathcal{O} \subset \mathbb{K}$ be the subring of elements with nonnegative valuation. Pick a uniformizer $\pi \in \mathbb{K}$ for the valuation v. We will assume that the residue field is finite.

Let $\underline{G}$ be a connected reductive group algebraic group over $\mathbb{K}$ and set $G = \underline{G}(\mathbb{K})$ and let $G^\vee = \underline{G}^\vee(\mathbb{C})$ be its Langlands dual. We will assume that $\underline{G}$ is split over

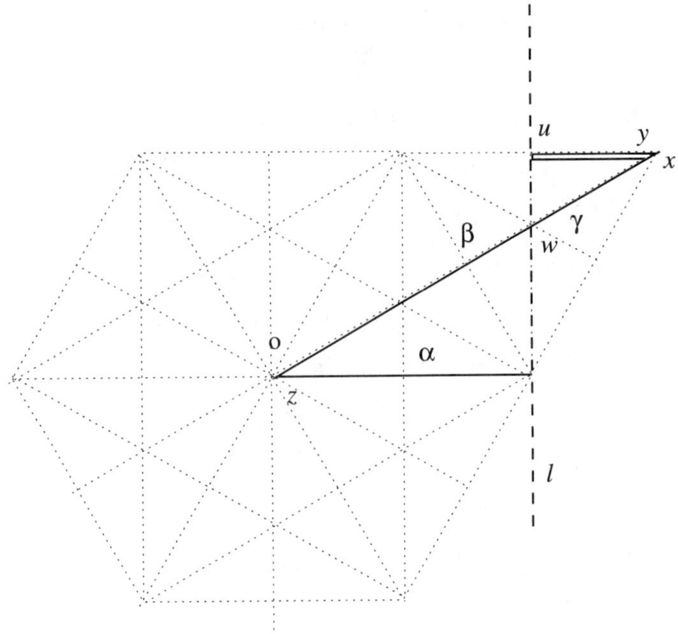

FIGURE 1. *A billiard triangle in the affine Coxeter complex of type G_2.*

$\mathbb{K}$. Recall that the *structure constants* $m_{\alpha,\beta}(\gamma)$ and $n_{\alpha,\beta}(\gamma)$ for the Hecke ring of G and the representation ring of $G^\vee$ respectively, are given by

$$c_\alpha \cdot c_\beta = \sum_\gamma m_{\alpha,\beta}(\gamma) c_\gamma,$$

$$ch(V_\alpha) \cdot ch(V_\beta) = \sum_\gamma n_{\alpha,\beta}(\gamma)\ ch(V_\gamma).$$

We will say that a dominant weight μ for $GL(\ell)$ is a *partition* if all components of the integer vector μ are nonnegative.

The starting point is that if the dominant weights α, β, γ are partitions then there is a standard formula [**Mac**, pg. 161] that expresses the structure constant $m_{\alpha,\beta}(\gamma)$ as the number of finite $\mathcal{O}$-module extensions

$$1 \to A \to B \to C \to 1,$$

where $A \cong \oplus_{i=1}^{\ell} \mathcal{O}/(\pi^{\alpha_i})$, $B \cong \oplus_{i=1}^{\ell} \mathcal{O}/(\pi^{\gamma_i})$ and $C \cong \oplus_{i=1}^{\ell} \mathcal{O}/(\pi^{\beta_i})$.

We will say that such an extension is of type (α, β, γ). Now for $GL(\ell)$ there is an explicit formula (the Littlewood–Richardson rule) for the structure constant $n_{\alpha,\beta}(\gamma)$ (in case α, β, γ are partitions) as the number of Littlewood–Richardson sequences (of partitions) of type (α, β, γ), see [**Mac**, pg. 68, 90]. Then the observation of J. Green is the following

LEMMA 8.26. *Suppose there exists an extension of $\mathcal{O}$-modules of type (α, β, γ). Then there exists a Littlewood-Richardson sequence (of partitions) of type (α, β, γ).*

Thus for the group $GL(\ell)$ we find that if α, β, γ^* are partitions and the structure constant $m_{\alpha,\beta}(\gamma^*)$ is nonzero then the structure constant $n_{\alpha,\beta}(\gamma^*)$ is nonzero (the superscript $*$ denotes the contragredient dominant weight).

We next remove the assumption that α, β, γ^* are partitions. To this end let λ be the cocharacter of $GL(\ell)$ that sends $z \in \mathbb{Q} \setminus \{0\}$ to the scalar matrix with diagonal entries equal to z. If α is a cocharacter then there exists k such that $\lambda^k \cdot \alpha$ is a partition. Now since the matrix $\lambda(\pi)$ is scalar it is clear that for any α $c_\alpha \cdot c_\lambda = c_{\alpha \cdot \lambda}$. It then follows that

$$m_{\lambda^k \cdot \alpha, \lambda^k \cdot \beta}(\lambda^{2k} \cdot \gamma^*) = m_{\alpha,\beta}(\gamma^*).$$

Now the previous operation of convolving with the function c_{λ_1} corresponds, under Langlands' duality, to multiplying a character by χ_1, the character of the determinant representation. Thus

$$m_{\alpha,\beta}(\gamma^*) \neq 0 \Leftrightarrow m_{\lambda^k \cdot \alpha, \lambda^k \cdot \beta}(\lambda^{2k} \cdot \gamma^*) \neq 0 \Leftrightarrow n_{\chi^k \cdot \alpha, \chi^k \cdot \beta}(\chi^{2k} \cdot \gamma^*) \neq 0 \Leftrightarrow n_{\alpha,\beta}(\gamma^*) \neq 0.$$

Finally, by Lemma 8.5 we have

$$m_{\alpha,\beta,\gamma}(0) \neq 0 \Leftrightarrow m_{\alpha,\beta}(\gamma^*) \neq 0.$$

Also it is immediate that $n_{\alpha,\beta}(\gamma^*) = n_{\alpha,\beta,\gamma}(0)$. Thus, if $(\alpha, \beta, \gamma) \in Sol(\mathbf{Q3})$ then $(\alpha, \beta, \gamma) \in Sol(\mathbf{Q4})$ as well. By combining this with Theorem 8.5, we conclude that for $GL(\ell)$ Problems **Q3** and **Q4** are equivalent. Since we have proved that 1 is a saturation factor for the Problem **Q3** it is also a saturation factor for **Q4**.

8.7. Computations for the root systems B_2 and G_2

Given a root system R, the intersection $D_3(X) \cap P(R^\vee)$ is a semigroup whose finite generating set (a *Hilbert basis*) can be computed once we know the explicit *stability inequalities* defining the convex cone $D_3(X)$. We have performed these computations for the root systems $R = B_2, G_2$ using the program 4ti2 (which could be found at http://www.4ti2.de) and the stability inequalities established in [**KLM1**]. Let ϖ_1, ϖ_2 be the long and short fundamental coweights for the root system R; below we will use the coordinates $[x, y]$ for the coweight $x\varpi_1 + y\varpi_2$.

B_2 **computation.** The Hilbert basis in the case $R = B_2$ consists of the following 8 triples (α, β, γ) and their permutations under the S_3 action:

$$\left(\begin{bmatrix} 0 \\ 1 \end{bmatrix}, \begin{bmatrix} 0 \\ 1 \end{bmatrix}, \begin{bmatrix} 0 \\ 0 \end{bmatrix}\right), \quad \left(\begin{bmatrix} 1 \\ 0 \end{bmatrix}, \begin{bmatrix} 1 \\ 0 \end{bmatrix}, \begin{bmatrix} 0 \\ 0 \end{bmatrix}\right),$$
$$\left(\begin{bmatrix} 1 \\ 0 \end{bmatrix}, \begin{bmatrix} 0 \\ 1 \end{bmatrix}, \begin{bmatrix} 0 \\ 1 \end{bmatrix}\right), \quad \left(\begin{bmatrix} 1 \\ 0 \end{bmatrix}, \begin{bmatrix} 1 \\ 0 \end{bmatrix}, \begin{bmatrix} 0 \\ 2 \end{bmatrix}\right),$$
$$\left(\begin{bmatrix} 1 \\ 0 \end{bmatrix}, \begin{bmatrix} 1 \\ 0 \end{bmatrix}, \begin{bmatrix} 1 \\ 0 \end{bmatrix}\right), \quad \left(\begin{bmatrix} 1 \\ 0 \end{bmatrix}, \begin{bmatrix} 1 \\ 0 \end{bmatrix}, \begin{bmatrix} 1 \\ 1 \end{bmatrix}\right),$$
$$\left(\begin{bmatrix} 0 \\ 1 \end{bmatrix}, \begin{bmatrix} 0 \\ 1 \end{bmatrix}, \begin{bmatrix} 0 \\ 1 \end{bmatrix}\right), \quad \left(\begin{bmatrix} 1 \\ 0 \end{bmatrix}, \begin{bmatrix} 1 \\ 0 \end{bmatrix}, \begin{bmatrix} 0 \\ 1 \end{bmatrix}\right).$$

We note that the first 5 generators are represented by flat triangles contained in an apartment in X and therefore are solutions of **Q4** (and of course of **Q3**). The last three generators are not solutions of **Q3** since $\alpha + \beta + \gamma \notin Q(R^\vee)$.

However, a direct computation (using Littelmann triangles) shows that for every generator (α, β, γ) among the last 3 generators in our list, $(2\alpha, 2\beta, 2\gamma)$ is a solution of **Q4**. Therefore, since the solution set of **Q4** forms a semigroup, we obtain

PROPOSITION 8.27. *If* (α,β,γ) *belongs to* $D_3(X)\cap X_*(\underline{T})$ *then* $(2\alpha,2\beta,2\gamma)$ *is a solution of the problem* **Q4** *for the group* $Sp(4)$.

We next observe that the solution set of the problem **Q3** for the group $Spin(5)$ is not a semigroup. Indeed, using the coordinates as in the Example 6.9, consider the triples

$$(\alpha'=(1,1),\beta'=(1,1),\gamma'=(1,1)),$$
$$(\alpha''=(1,1),\beta''=(1,1),\gamma''=(2,0)).$$

As in section 8.5, the first triple is a solution of the problem **Q3** for $Spin(5)$, the second triple represents a flat triangle with the vertices $(0,0),(1,1),(2,0)$ in the apartment. However

$$(\alpha'+\alpha'',\beta'+\beta'',\gamma'+\gamma'')=(\alpha,\beta,\gamma)$$

is the triple from the Example 6.9, and therefore is not a solution of **Q3**.

G_2 **computation.** The Hilbert basis in the case $R=G_2$ consists of the following triples (α,β,γ) and their permutations under the S_3 action:

$$\left(\begin{bmatrix}0\\1\end{bmatrix},\begin{bmatrix}0\\1\end{bmatrix},\begin{bmatrix}0\\0\end{bmatrix}\right),\quad \left(\begin{bmatrix}1\\0\end{bmatrix},\begin{bmatrix}1\\0\end{bmatrix},\begin{bmatrix}0\\0\end{bmatrix}\right),$$
$$\left(\begin{bmatrix}0\\1\end{bmatrix},\begin{bmatrix}0\\1\end{bmatrix},\begin{bmatrix}0\\1\end{bmatrix}\right),\quad \left(\begin{bmatrix}1\\0\end{bmatrix},\begin{bmatrix}1\\0\end{bmatrix},\begin{bmatrix}1\\0\end{bmatrix}\right),$$
$$\left(\begin{bmatrix}1\\0\end{bmatrix},\begin{bmatrix}1\\0\end{bmatrix},\begin{bmatrix}0\\3\end{bmatrix}\right),\quad \left(\begin{bmatrix}1\\0\end{bmatrix},\begin{bmatrix}2\\0\end{bmatrix},\begin{bmatrix}0\\3\end{bmatrix}\right),$$
$$\left(\begin{bmatrix}1\\0\end{bmatrix},\begin{bmatrix}0\\1\end{bmatrix},\begin{bmatrix}0\\1\end{bmatrix}\right),\quad \left(\begin{bmatrix}1\\0\end{bmatrix},\begin{bmatrix}0\\1\end{bmatrix},\begin{bmatrix}0\\2\end{bmatrix}\right),$$
$$\left(\begin{bmatrix}1\\0\end{bmatrix},\begin{bmatrix}1\\0\end{bmatrix},\begin{bmatrix}0\\2\end{bmatrix}\right),$$
$$\left(\begin{bmatrix}1\\0\end{bmatrix},\begin{bmatrix}1\\0\end{bmatrix},\begin{bmatrix}0\\1\end{bmatrix}\right),\quad \left(\begin{bmatrix}1\\0\end{bmatrix},\begin{bmatrix}1\\0\end{bmatrix},\begin{bmatrix}1\\1\end{bmatrix}\right).$$

The first 9 generators are solutions of **Q4** and the last two generators are only solutions of **Q3**. However one can show (arguing analogously to the Example 6.9) that the sum of the last two generators,

$$\left(\begin{bmatrix}2\\0\end{bmatrix},\begin{bmatrix}2\\0\end{bmatrix},\begin{bmatrix}1\\2\end{bmatrix}\right)$$

is not a solution of **Q3**. Hence the solution set of **Q3** for the root system G_2 is not a semigroup. On the other hand, a direct computation shows that for both $k=2,k=3$ the triples

$$k\cdot\left(\begin{bmatrix}1\\0\end{bmatrix},\begin{bmatrix}1\\0\end{bmatrix},\begin{bmatrix}0\\1\end{bmatrix}\right),\quad k\cdot\left(\begin{bmatrix}1\\0\end{bmatrix},\begin{bmatrix}1\\0\end{bmatrix},\begin{bmatrix}1\\1\end{bmatrix}\right)$$

are solutions of **Q4**. Therefore, since the solution set of **Q4** is a semigroup and because each natural number $k\geq 2$ has the form $2n+3m, n,m\in\mathbb{N}\cup\{0\}$, we get:

PROPOSITION 8.28. *For each* $k\in\mathbb{N}\setminus\{1\}$ *the semigroup*

$$k\cdot(D_3(X)\cap P(R^\vee))$$

is contained in the solution set of **Q4**.

APPENDIX A

Decomposition of Tensor Products and Mumford Quotients of Products of Coadjoint orbits

A.1. The existence of semistable triples and nonzero invariant vectorsin triple tensor products

In this section we will assume that G is a reductive complex Lie group with Lie algebra $\mathfrak{g}$ and Weyl group W. We let K be a maximal compact subgroup of G and T be a maximal torus in G and B be a Borel subgroup containing T. We let $\mathfrak{h}$ denote the Lie algebra of T. We may choose T so that it is preserved by the Cartan involution of G corresponding to K. Let $\mathfrak{g} = \mathfrak{k} \oplus \mathfrak{p}$ denote the Cartan decomposition of the Lie algebra $\mathfrak{g}$ of G and let $\mathfrak{a} := \mathfrak{h} \cap \mathfrak{p}$. Then multiplication by $\sqrt{-1}$ interchanges $\mathfrak{p}$ and $\mathfrak{k}$. We let $X := G/K$ be the associated symmetric space. The choice of B is equivalent to a choice of positive roots or a positive chamber $\Delta \subset \mathfrak{a}$. We will be sloppy throughout with the difference between the solution sets for Problem **Q1** for G and $G^\vee$ since these sets are canonically isomorphic (say by using an $Ad\, K$ invariant metric on $\mathfrak{k}$). In the similar fashion we will identify $\mathfrak{k}$ with $\mathfrak{k}^*$ and $\mathfrak{p}$ with $\mathfrak{p}^*$.

We will use the following notation. Let $\lambda \in \mathfrak{a}^*$ be a dominant weight for the torus T. Then we let V_λ be an irreducible G–module with the highest weight λ (so V_λ is unique up to isomorphism). We let λ^* be the highest weight of the dual representation, so $\lambda^* = w_0(-\lambda)$ where w_0 is the longest element in W.

For each fundamental weight λ belonging to $\mathfrak{a}^* \subset \mathfrak{p}^*$ we define the *coadjoint orbit*, $\mathcal{O}_\lambda \subset \mathfrak{k}^*$, to be the Ad^*K–orbit of $\sqrt{-1}\lambda$.

Let $\lambda \in \mathfrak{a}^*$ be a dominant weight. Then the coadjoint orbit $\mathcal{O}_\lambda$ as above carries a natural homogeneous complex structure (see [**Vo**, Chapter 1]) so that the coadjoint action of K on $\mathcal{O}_\lambda$ extends to a holomorphic action of G. The dominant weight λ defines a very ample line bundle $\mathbb{L}_\lambda$ over $\mathcal{O}_\lambda$ and the action of G on $\mathcal{O}_\lambda$ extends to a holomorphic action $G \curvearrowright \mathbb{L}_\lambda$.

The key point in what follows is the famous theorem of Borel and Weil below, Theorem A.1, see e.g. [**Vo**, Chapter 1].

THEOREM A.1. *The space of holomorphic sections $\Gamma(\mathcal{O}_\lambda, \mathbb{L}_\lambda)$ of $\mathbb{L}_\lambda$ is isomorphic as a G–module to the G–module V_λ.*

Now let α and β be dominant weights. In this section we will discuss the problem of finding the possible irreducible constituents of tensor products $V_\alpha \otimes V_\beta$. Of course this is equivalent to finding for which triples α, β, γ, the space of G–invariants

$$(V_\alpha \otimes V_\beta \otimes V_\gamma)^G \cong Hom_G(V_{\gamma^*}, V_\alpha \otimes V_\beta)$$

is nonzero.

Given dominant weights α, β, γ we define complex G–manifold

$$\mathcal{O}_{\alpha,\beta,\gamma} := \mathcal{O}_\alpha \times \mathcal{O}_\beta \times \mathcal{O}_\gamma.$$

We also have the outer tensor product

$$\mathbb{L}_{\alpha,\beta,\gamma} := \mathbb{L}_\alpha \boxtimes \mathbb{L}_\beta \boxtimes \mathbb{L}_\gamma$$

which is a very ample G–invariant line bundle over $\mathcal{O}_{\alpha,\beta,\gamma}$.

Since the line bundle $\mathbb{L}_{\alpha,\beta,\gamma}$ is very ample, it determines a G—equivariant holomorphic embedding F of $\mathcal{O}_{\alpha,\beta,\gamma}$ into $\mathbb{P}((V_\alpha \otimes V_\beta \otimes V_\gamma)^*)$. We will use the notation

$$\mathcal{M}_{\alpha,\beta,\gamma} := F(\mathcal{O}_{\alpha,\beta,\gamma})//G$$

for the Mumford quotient associated with this line bundle. See the end of section 5.1 for further discussion; note that here we have shortened the notation $\mathcal{M}_{(\alpha,\beta,\gamma),sst}(B)$ (in §5.1) to $\mathcal{M}_{\alpha,\beta,\gamma}$.

LEMMA A.1. *The moduli space of triangles in the infinitesimal symmetric space $\mathfrak{p}$ with Δ–side lengths α, β, γ is canonically homeomorphic to the Mumford quotient $\mathcal{M}_{\alpha,\beta,\gamma}$ defined above.*

Proof: By a theorem of Kempf and Ness [**KN**], the Mumford quotient $\mathcal{M}_{\alpha,\beta,\gamma}$ is canonically homeomorphic to the symplectic quotient $\mathcal{O}_{\alpha,\beta,\gamma}//Ad^*(K)$. It is a standard argument (essentially the formula for the moment map of the action of the diagonal subgroup of a product) that

$$\mathcal{O}_{\alpha,\beta,\gamma}//Ad^*(K) = \{(\lambda, \mu, \nu) \in \mathcal{O}_{\alpha,\beta,\gamma} : \lambda + \mu + \nu = 0\}/Ad^*(K).$$

However the latter is canonically isomorphic to the moduli space of triangles in $\mathfrak{p}$ with the Δ–side lengths $\alpha, \beta, \gamma \in \Delta$. □

Thus

$$\mathcal{M}_{\alpha,\beta,\gamma} \neq \emptyset \iff (\alpha, \beta, \gamma) \in D_3(\mathfrak{p}) \iff (\alpha, \beta, \gamma) \in D_3(X).$$

The main goal of this section is to prove Theorem A.2, that relates the existence of semistable triples in $\mathcal{O}_{\alpha,\beta,\gamma}$ and the existence of nonzero invariants in triple tensor products of irreducible representations of G.

THEOREM A.2. *For each triple of dominant weights $\alpha, \beta, \gamma \in \mathfrak{a}^*$ the following are equivalent:*

1. *There exists a positive integer $k > 0$ such that $(V_{k\alpha} \otimes V_{k\beta} \otimes V_{k\gamma})^G \neq 0$.*
2. *$\mathcal{M}_{\alpha,\beta,\gamma} \neq \emptyset$, i.e. there exists a semistable point in $\mathcal{O}_{\alpha,\beta,\gamma}$.*

REMARK A.2. Theorem A.2 implies that if $(V_\alpha \otimes V_\beta \otimes V_\gamma)^G \neq 0$ then there exists a weighted semistable configuration on $\partial_{Tits}X$ of type (α, β, γ). However the converse is false as we have seen in section 8.5.

Theorem A.2 will follow from the next two lemmas.

Let A be the graded ring associated to the projective variety $F(\mathcal{O}_{\alpha,\beta,\gamma})$. Then by definition $A^{(k)}$ is the set of restrictions to $F(\mathcal{O}_{\alpha,\beta,\gamma})$ of the homogeneous polynomials of degree k in the projective coordinates given by a basis of

$$V_\alpha \otimes V_\beta \otimes V_\gamma = ((V_\alpha \otimes V_\beta \otimes V_\gamma)^*)^*.$$

Thus

$$A^{(k)} = \Gamma(\mathcal{O}_{\alpha,\beta,\gamma}, \mathbb{L}_{\alpha,\beta,\gamma})^{\otimes k}/I^{(k)}.$$

Here $I^{(k)}$ is the degree k–summand of the graded ideal of polynomials in the sections of $\mathbb{L}_{\alpha,\beta,\gamma}$ that vanish on $\mathcal{O}_{\alpha,\beta,\gamma}$. Equivalently, $I^{(k)}$ is the degree k component of the ideal I of polynomials that vanish on $F(\mathcal{O}_{\alpha,\beta,\gamma})$ in the ring of polynomials on $\mathbb{P}((V_\alpha \otimes V_\beta \otimes V_\gamma)^*)$.

Now another definition of the Mumford quotient is $Proj(A^G)$ where A^G is the subring of G–invariants of the graded ring A. We define another graded ring R by

$$R = \bigoplus_{k=0}^{\infty} \Gamma(\mathcal{O}_{\alpha,\beta,\gamma}, (\mathbb{L}_\alpha \boxtimes \mathbb{L}_\beta \boxtimes \mathbb{L}_\gamma)^{\otimes k}).$$

Below and in what follows, if U_1, U_2 are G_1, G_2–modules, then $U_1 \boxtimes U_2$ denotes the $G_1 \times G_2$–module with the underlying vector space equal to the tensor product $U_1 \otimes U_2$.

We define a graded $G \times G \times G$–module R' by

$$R' = \bigoplus_{k=0}^{\infty} (V_{k\alpha} \boxtimes V_{k\beta} \boxtimes V_{k\gamma}).$$

We will abuse notation and use R' to denote the restriction of the previous module to the diagonal in $G \times G \times G$. Then, we have an isomorphism of G–modules

$$R' = \bigoplus_{k=0}^{\infty} (V_{k\alpha} \otimes V_{k\beta} \otimes V_{k\gamma}).$$

LEMMA A.3. *There is a canonical isomorphism of graded $(G \times G \times G)$–modules $R \cong R'$.*

Proof: The lemma follows from the following three equations. First, let M_1 and M_2 be complex manifolds and $\mathbb{L}_1$, resp. $\mathbb{L}_2$, be a holomorphic line bundle over M_1, resp. M_2. First we have

$$(\mathbb{L}_1 \boxtimes \mathbb{L}_2)^{\otimes k} = \mathbb{L}_1^{\otimes k} \boxtimes \mathbb{L}_2^{\otimes k}.$$

Next, we have

$$\Gamma(M_1 \times M_2, \mathbb{L}_1 \boxtimes \mathbb{L}_2) \cong \Gamma(M_1, \mathbb{L}_1) \otimes \Gamma(M_2, \mathbb{L}_2).$$

Finally, for a natural number k, a dominant weight α and the corresponding orbit $\mathcal{O}_\alpha$ we have

$$\Gamma(\mathcal{O}_\alpha, \mathbb{L}_\alpha^{\otimes k}) \cong \Gamma(\mathcal{O}_\alpha, \mathbb{L}_{k\alpha}). \quad \square$$

Next we show that A and R are isomorphic. This is the exceptional feature of the homogeneous situation. The key point, the surjectivity of the natural map below, was pointed out to us by Lawrence Ein.

LEMMA A.4. *There is a natural $G \times G \times G$–equivariant isomorphism from the graded ring A to the graded ring R.*

Proof: To obtain the desired map from the graded ring A to the graded ring R we observe that for any complex manifold M and holomorphic line bundle $\mathbb{L}$ over M there is a natural map (usually not onto) from $\Gamma(M, \mathbb{L})^{\otimes k}$ to $\Gamma(M, \mathbb{L}^{\otimes k})$. Hence there is a natural map from $A^{(k)}$ to $R^{(k)}$. The exceptional feature here is that this natural map is *onto*.

To see this we note that by the theorem of Borel and Weil the action of $G\times G\times G$ on $R^{(k)}$ is irreducible. But the image of $A^{(k)}$ in $R^{(k)}$ is an invariant subspace. Hence $A^{(k)}$ maps onto $R^{(k)}$. Clearly the map is injective (because we have divided by the ideal $I^{(k)}$). □

Theorem A.2 follows by taking G–invariants from the G–isomorphism $A \cong R'$. Indeed,

$$\text{There exists a triangle with } \Delta\text{–side lengths } \alpha, \beta, \gamma \Leftrightarrow$$

$$\text{The Mumford quotient } \mathcal{M}_{\alpha,\beta,\gamma} \text{ is nonempty } \Leftrightarrow$$

$$(\bigoplus_{k=1}^{\infty} A^{(k)})^G \neq 0 \Leftrightarrow$$

$$\bigoplus_{k=1}^{\infty}(V_{k\alpha} \otimes V_{k\beta} \otimes V_{k\gamma})^G \neq 0.$$

□

REMARK A.5. The reason that the existence of a nonzero G–invariant does not follow from the existence of a triangle is that the lowest degree G–invariant might not have degree 1, that is, it is possible that

$$(R^{(1)})^G = \{0\} \text{ but } (R^{(k)})^G \neq 0 \text{ for some } k > 1.$$

A.2. The semigroups of solutions to Problems Q1 and Q4

We will use S_G to denote the semigroup of dominant characters of T henceforth. We begin this section by noting that the set of triples of dominant characters (α, β, γ) that belong to $D_3(X)$ is a subsemigroup $S_{triangle} \subset S_G^3$ (since it is determined by a system of homogeneous linear inequalities). Moreover, since these inequalities have integral coefficients, the subsemigroup $S_{triangle}$ is finitely generated. We set

$$S_{rep} := Sol(\mathbf{Q4}, G) \subset D_3(X) \cap S_G^3,$$

i.e. the set of triples (α, β, γ) for which

$$(V_\alpha \otimes V_\beta \otimes V_\gamma)^G \neq 0.$$

We recall that if $S_1 \subset S_2$ is an inclusion of semigroups then the *saturation* of S_1 in S_2 is the semigroup of elements of $s \in S_2$ such that for some positive integer n we have $ns \in S_1$.

THEOREM A.3. (1) *The set S_{rep} is a subsemigroup of the semigroup S_G^3 (and of $S_{triangle}$).*
(2) *The saturation of S_{rep} in S_G^3 is $S_{triangle}$.*

First we need a general lemma. Let G_1 be a complex reductive group, B_1 be a Borel subgroup of G_1 and λ and μ be dominant characters (as in the beginning of the appendix). We recall that the irreducible representation $V_{\lambda+\mu}$ is always an irreducible constituent of multiplicity 1 in the tensor product $V_\lambda \otimes V_\mu$. Let $\pi : V_\lambda \otimes V_\mu \to V_{\lambda+\mu}$ be the G_1–equivariant projection.

LEMMA A.6. *Let $v_1 \in V_\lambda$ and $v_2 \in V_\mu$ be nonzero vectors. Then*

$$\pi(v_1 \otimes v_2) \neq 0.$$

Proof: Let $M = G_1/B_1$. We apply the Borel-Weil Theorem to obtain $V_\lambda = \Gamma(M, \mathbb{L}_\lambda)$, $V_\mu = \Gamma(M, \mathbb{L}_\mu)$ and $V_{\lambda+\mu} = \Gamma(M, \mathbb{L}_{\lambda+\mu})$. Then v_1 corresponds to a section s_1 and v_2 corresponds to a section s_2. Hence $\pi(v_1 \otimes v_2)$ corresponds to the product section $s_1 \cdot s_2$ of the product line bundle $\mathbb{L}_{\lambda+\mu}$. But since M is irreducible the product section cannot be zero. □

Now we can prove the theorem.

Proof: Let $(\alpha_i, \beta_i, \gamma_i) \in S_{rep}$ and let v_i be a nonzero G–invariant vector in

$$V_{\alpha_i} \otimes V_{\beta_i} \otimes V_{\gamma_i}, i = 1, 2.$$

We take $G_1 := G \times G \times G$ and $B_1 := B \times B \times B$, where $B \subset G$ is a Borel subgroup as before; we have the G_1–modules

$$V_\lambda := V_{\alpha_1} \boxtimes V_{\beta_1} \boxtimes V_{\gamma_1}, V_\mu := V_{\alpha_2} \boxtimes V_{\beta_2} \boxtimes V_{\gamma_2}.$$

Then

$$V_\lambda \otimes V_\mu \cong (V_{\alpha_1} \otimes V_{\alpha_2}) \boxtimes (V_{\beta_1} \otimes V_{\beta_2}) \boxtimes (V_{\gamma_1} \otimes V_{\gamma_2}).$$

By the above lemma, the vector $\pi(v_1 \otimes v_2)$ is nonzero. Since it is clearly G–invariant,

$$(V_{\alpha_1+\alpha_2} \boxtimes V_{\beta_1+\beta_2} \boxtimes V_{\gamma_1+\gamma_2})^G \neq 0.$$

Hence $(\alpha_1 + \alpha_2, \beta_1 + \beta_2, \gamma_1 + \gamma_2) \in S_G$ and the first statement of Theorem follows.

The second statement follows from Theorem A.2. □

An alternative proof of Part 1 of Theorem A.3 follows from the description of S_{rep} given in [**BZ**] by Berenstein and Zelevinsky.

THEOREM A.4. *There exists a (nonzero) $k \in \mathbb{N}$ depending only on the group G so that $k \cdot S_{triangle} \subset S_{rep}$.*

Proof: By Gordan's lemma, see [**Fu2**, Proposition 1, Page 12], the semigroup $S_{triangle}$ is finitely generated. Choose a finite set of generators $s_1, ..., s_m$ of this semigroup. For each generator s_i there exists a positive integer k_i such that

$$k_i s_i \in S_{rep},$$

see Theorem A.2. Then take $k = LCM(k_1, ..., k_m)$. □

Therefore, since S_{rep} contains finitely generated semigroup $k \cdot S_{triangle}$ as a subsemigroup of finite index, we get

COROLLARY A.7. *S_{rep} is finitely generated.*

Bibliography

[AW] S. Agnihotri and C. Woodward, *Eigenvalues of products of unitary matrices and quantum Schubert calculus*, Math. Res. Letters, vol. **5** (1998), p. 817–836.

[AMW] A. Alekseev, E. Meinrenken and C. Woodward, *Linearization of Poisson actions and singular values of matrix products*, Ann. Inst. Fourier (Grenoble), vol. **51** (2001), no. 6, p. 1691–1717.

[Ba] W. Ballmann, "Lectures on spaces of nonpositive curvature. With an appendix by Misha Brin." DMV Seminar, vol. 25. Birkhauser Verlag, Basel, 1995.

[B] A. Balser, *Polygons with prescribed Gauss map in Hadamard spaces and Euclidean buildings*, Canadian Mathematical Bulletin, to appear.

[Bel] P. Belkale, *Local systems on* $\mathbb{P}^1$*-*S *for* S *a finite set*, Compositio Math., vol. **129** (2001) p. 67–86.

[BeSj] A. Berenstein and R. Sjamaar, *Coadjoint orbits, moment polytopes, and the Hilbert-Mumford criterion*, Journ. Amer. Math. Soc., vol. **13** (2000), no. 2, p. 433–466.

[BZ] A. Berenstein and A. Zelevinsky, *Tensor product multiplicities, canonical bases and totally positive varieties*, Invent. Math., vol. **143** (2001), no. 1, p. 77–128.

[BK] P. Belkale, S. Kumar, *Eigenvalue problem and a new product in cohomology of flag varieties*, Preprint, math.AG/0407034, 2004.

[Bo] N. Bourbaki, "Lie groups and Lie algebras", Chap. 4, 5, 6. Springer Verlag, 2002.

[BT] F. Bruhat and J. Tits, *Groupes reductifs sur un corps local,* Inst. Hautes Études Sci. Publ. Math., No. 41, (1972), p. 5–251.

[C] P. Cartier, *Representations of p-adic groups*, In: "Automorphic forms, representations, and L-functions", Proc. AMS, vol. **33** (1979), p. 111–155.

[DW] H. Derksen, J. Weyman, *Semi-invariants of quivers and saturation for Littlewood–Richardson coefficients*, J. Amer. Math. Soc., vol. **13**, (2000), n. 3, p. 467–479.

[EL] S. Evens, J.-H. Lu, *Thompson's conjecture for real semi-simple Lie groups*, Preprint, ArXiv, math.SG/0310098.

[Fu1] W. Fulton, *Eigenvalues, invariant factors, highest weights, and Schubert calculus*, Bull. Amer. Math. Soc. (N.S.), vol. **37** (2000), no. 3, p. 209–249.

[Fu2] W. Fulton, "Introduction to toric varieties", Annals of Math. Studies, 131, Princeton Univerity Press, 1993.

[FH] W. Fulton, J. Harris, "Representation theory. A first course", Graduate Texts in Math. , vol. **129**, Springer Verlag, 1991.

[G] F. Gauss, *Letter to W. Bolyai*, "Collected Works", vol. 8, p. 222–223.

[Gro] B. Gross, *On the Satake isomorphism,* In: "Galois representations in arithmetic algebraic geometry (Durham, 1996)", London Math. Soc. Lecture Notes, vol. **254**, (1998) p. 223–237.

[HL] P. Heinzner and F. Loose, *Reduction of complex Hamiltonian G-spaces*, Geom. Funct. Anal., vol. **4** (1994), no. 3, p. 288–297.

[Ha1] T. J. Haines, *On the matrix coefficients of the Satake isomorphism: complements to the paper of M. Rapoport*, Manuscripta Math., vol. **101** (2000), p. 167–174.

[Ha2] T. J. Haines, *Structure constants for Hecke and representations rings*, IMRN vol. **39** (2003), p. 2103–2119.

[Hum] J. Humphreys, "Reflection groups and Coxeter groups", Cambridge University Press, 1990.

[KM1] M. Kapovich, J. J. Millson, *The symplectic geometry of polygons in Euclidean space*, Journal of Diff. Geometry, vol. **44**, (1996), p. 479–513.

[KM2] M. Kapovich, J. J. Millson, *A path model for geodesic in Euclidean buildings and its applications to the representation theory*, Preprint, November 2004.

[KLM1] M. Kapovich, B. Leeb, J. J. Millson, *Convex functions on symmetric spaces, side lengths of polygons and the stability inequalities for weighted configurations at infinity*, Preprint, June 2004.

[KLM2] M. Kapovich, B. Leeb, J. J. Millson, *Polygons in buildings and their refined side-lengths*, Preprint, June 2004.

[KMT] M. Kapovich, J. J. Millson and T. Treloar, *The symplectic geometry of polygons in hyperbolic 3-space,* Asian J. Math. (Kodaira's issue), vol. **4** (2000), no. 1, p. 123–164.

[KN] G. Kempf and L. Ness, *The length of vectors in representation spaces,* In: "Algebraic Geometry Proceedings. Copenhagen 1978", Springer Lecture Notes in Mathematics, vol. **732** (1979), p. 233–243.

[Kle1] T. Klein, *The multiplication of Schur functions and extensions of p-modules*, London Math. Soc. J., vol **43** (1968), p. 280–284.

[Kle2] T. Klein, *The Hall polynomial*, J. Algebra, vol. **12** (1969), p. 61–78.

[KL] B. Kleiner, B. Leeb, *Rigidity of quasi-isometries for symmetric spaces and Euclidean buildings*, Publ. Math. IHES, vol. **86** (1997), p. 115–197.

[Kly1] A. Klyachko, *Stable bundles, representation theory and Hermitian operators*, Selecta Mathematica, vol. **4** (1998), p. 419–445.

[Kly2] A. Klyachko, *Random walks on symmetric spaces and inequalities for matrix spectra*, Linear Algebra Appl. Special Issue: Workshop on Geometric and Combinatorial Methods in the Hermitian Sum Spectral Problem (Coimbra, 1999), vol. **319** (2000), no. 1–3, p. 37–59.

[KT] A. Knutson and T. Tao, *The honeycomb model of* $\mathrm{GL}_n(\mathbb{C})$ *tensor products. I. Proof of the saturation conjecture,* J. Amer. Math. Soc., vol. **12** (1999), no. 4, p. 1055–1090.

[Ko] S. Kochen, *The model theory of local fields*, ISILC Logic conference (Proc. Internat. Summer Inst. and Logic Colloq., Kiel, 1974), Springer Lecture Notes in Math., vol. **499**, p. 384–425.

[KuLM] S. Kumar, B. Leeb, J. J. Millson, *The generalized triangle inequalities for rank* 3 *symmetric space of noncompact type*, In: "Explorations in complex and Riemannian geometry (Papers dedicated to Robert Greene)," Contemporary Math., Vol 332, 2003, p. 171–195.

[La] M. C. Laskowski, *An application of Kochen's Theorem,* J. Symbolic Logic, **68** (2003), no. 4, 1181–1188.

[L1] P. Littelmann, *Generalization of the Littlewood–Richardson Rule*, Invent. Math., vol. **116** (1994), p. 329–346.

[L2] P. Littelmann, *Paths and root operators in representation theory*, Annals of Math. (2) 142 (1995) no. 3, p. 499–525.

[Lu1] G. Lusztig, *Green polynomials and singularities of unipotent classes*, Adv. in Math., vol. **42** (1981), p. 169–178.

[Lu2] G. Lusztig, *Singularities, character formulas, and a q-analogue of weight multiplicities*, Astérisque, vol. **101** (1983), p. 208–227.

[Mac] I. G. Macdonald, *Symmetric functions and Hall polynomials*, Oxford Mathematical Monographs, Clarendon Press, Oxford, 1979.

[Sam] H. Samelson, *Notes on Lie Algebras*, Van Nostrand, 1969.

[Sat] I. Satake, *Theory of spherical functions on reductive algebraic groups over p-adic fields*, Publ. Math. IHES, vol. **18** (1963), p. 1–69.

[Sj] R. Sjamaar, *Holomorphic slices, symplectic reduction and multiplicities of representations*, Ann. of Math., vol. **141** (1995), no. 1, p. 87–129.

[Sp] T. A. Springer, *Reductive groups* , In: "Automorphic forms, representations and L-functions", Proc. Symp. AMS, vol. **33** (1979), p. 3–27.

[SpSt] T. A. Springer and R. Steinberg, *Conjugacy classes*, In: "Seminar on Algebraic Groups and Related Finite Groups", Springer Lecture Notes in Math., vol. **131** (1970), p. 167–266.

[Ta] M. H. Taibleson, "Fourier Analysis on Local Fields", Mathematical Notes, Vol. **15**, Princeton University Press, 1975.

[Tits] J. Tits, *Reductive groups over local fields*, In: "Automorphic forms, representations and L-functions", Proc. Symp. AMS, vol. **33** (1979), p. 29–69.

[Tr] T. Treloar, *The symplectic geometry of polygons in the 3-sphere*, Canad. J. Math., vol. **54**, (2002), p. 30–54.

[Vo] D. Vogan,"Unitary representations of reductive Lie groups", Princeton University Press, 1987.

Editorial Information

To be published in the *Memoirs*, a paper must be correct, new, nontrivial, and significant. Further, it must be well written and of interest to a substantial number of mathematicians. Piecemeal results, such as an inconclusive step toward an unproved major theorem or a minor variation on a known result, are in general not acceptable for publication.

Papers appearing in *Memoirs* are generally at least 80 and not more than 200 published pages in length. Papers less than 80 or more than 200 published pages require the approval of the Managing Editor of the Transactions/Memoirs Editorial Board.

As of November 30, 2007, the backlog for this journal was approximately 14 volumes. This estimate is the result of dividing the number of manuscripts for this journal in the Providence office that have not yet gone to the printer on the above date by the average number of monographs per volume over the previous twelve months, reduced by the number of volumes published in four months (the time necessary for preparing a volume for the printer). (There are 6 volumes per year, each usually containing at least 4 numbers.)

A Consent to Publish and Copyright Agreement is required before a paper will be published in the *Memoirs*. After a paper is accepted for publication, the Providence office will send a Consent to Publish and Copyright Agreement to all authors of the paper. By submitting a paper to the *Memoirs*, authors certify that the results have not been submitted to nor are they under consideration for publication by another journal, conference proceedings, or similar publication.

Information for Authors

Memoirs are printed from camera copy fully prepared by the author. This means that the finished book will look exactly like the copy submitted.

Initial submission. The AMS uses Centralized Manuscript Processing for initial submissions. Authors should submit a PDF file using the Initial Manuscript Submission form found at `www.ams.org/cgi-bin/peertrack/submission.pl`, or send one copy of the manuscript to the following address: Centralized Manuscript Processing, MEMOIRS OF THE AMS, 201 Charles Street, Providence, RI 02904-2294 USA. If a paper copy is being forwarded to the AMS, indicate that it is for it Memoirs and include the name of the corresponding author, contact information such as email address or mailing address, and the name of an appropriate Editor to review the paper (see the list of Editors below).

The paper must contain a *descriptive title* and an *abstract* that summarizes the article in language suitable for workers in the general field (algebra, analysis, etc.). The *descriptive title* should be short, but informative; useless or vague phrases such as "some remarks about" or "concerning" should be avoided. The *abstract* should be at least one complete sentence, and at most 300 words. Included with the footnotes to the paper should be the 2000 *Mathematics Subject Classification* representing the primary and secondary subjects of the article. The classifications are accessible from `www.ams.org/msc/`. The list of classifications is also available in print starting with the 1999 annual index of *Mathematical Reviews*. The Mathematics Subject Classification footnote may be followed by a list of *key words and phrases* describing the subject matter of the article and taken from it. Journal abbreviations used in bibliographies are listed in the latest *Mathematical Reviews* annual index. The series abbreviations are also accessible from `www.ams.org/publications/`. To help in preparing and verifying references, the AMS offers MR Lookup, a Reference Tool for Linking, at `www.ams.org/mrlookup/`.

Electronically prepared manuscripts. The AMS encourages electronically prepared manuscripts, with a strong preference for $\mathcal{A}_{\mathcal{M}}\mathcal{S}$-LaTeX. To this end, the Society has prepared $\mathcal{A}_{\mathcal{M}}\mathcal{S}$-LaTeX author packages for each AMS publication. Author packages include instructions for preparing electronic manuscripts, samples, and a style file that generates the particular design specifications of that publication series. Though $\mathcal{A}\mathcal{M}\mathcal{S}$-LaTeX is the highly preferred format of TeX, author packages are also available in $\mathcal{A}\mathcal{M}\mathcal{S}$-TeX.

Authors may retrieve an author package from the AMS website starting from `www.ams.org/tex/` or via FTP to `ftp.ams.org` (login as `anonymous`, enter username as password, and type `cd pub/author-info`). The *AMS Author Handbook* and the *Instruction Manual* are available in PDF format following the author packages link from `www.ams.org/tex/`. The author package can also be obtained free of charge by sending email to `tech-support@ams.org` (Internet) or from the Publication Division, American Mathematical Society, 201 Charles St., Providence, RI 02904-2294, USA. When requesting an author package, please specify $\mathcal{A}\mathcal{M}\mathcal{S}$-LaTeX or $\mathcal{A}\mathcal{M}\mathcal{S}$-TeX and the publication in which your paper will appear. Please be sure to include your complete mailing address.

After acceptance. The final version of the electronic file should be sent to the Providence office (this includes any TeX source file, any graphics files, and the DVI or PostScript file) immediately after the paper has been accepted for publication.

Before sending the source file, be sure you have proofread your paper carefully. The files you send must be the EXACT files used to generate the proof copy that was accepted for publication. For all publications, authors are required to send a printed copy of their paper, which exactly matches the copy approved for publication, along with any graphics that will appear in the paper.

Accepted electronically prepared files can be submitted via the web at `www.ams.org/submit-book-journal/`, sent via FTP, or sent on CD-Rom or diskette to the Electronic Prepress Department, American Mathematical Society, 201 Charles Street, Providence, RI 02904-2294 USA. TeX source files, DVI files, and PostScript files can be transferred over the Internet by FTP to the Internet node `ftp.ams.org` (130.44.1.100). When sending a manuscript electronically via CD-Rom or diskette, please be sure to include a message identifying the paper as a Memoir.

Electronically prepared manuscripts can also be sent via email to `pub-submit@ams.org` (Internet). In order to send files via email, they must be encoded properly. (DVI files are binary and PostScript files tend to be very large.)

Electronic graphics. Comprehensive instructions on preparing graphics are available at `www.ams.org/jourhtml/`. A few of the major requirements are given here.

Submit files for graphics as EPS (Encapsulated PostScript) files. This includes graphics originated via a graphics application as well as scanned photographs or other computer-generated images. If this is not possible, TIFF files are acceptable as long as they can be opened in Adobe Photoshop or Illustrator. No matter what method was used to produce the graphic, it is necessary to provide a paper copy to the AMS.

Authors using graphics packages for the creation of electronic art should also avoid the use of any lines thinner than 0.5 points in width. Many graphics packages allow the user to specify a "hairline" for a very thin line. Hairlines often look acceptable when proofed on a typical laser printer. However, when produced on a high-resolution laser imagesetter, hairlines become nearly invisible and will be lost entirely in the final printing process.

Screens should be set to values between 15% and 85%. Screens which fall outside of this range are too light or too dark to print correctly. Variations of screens within a graphic should be no less than 10%.

Inquiries. Any inquiries concerning a paper that has been accepted for publication should be sent to `memo-query@ams.org` or directly to the Electronic Prepress Department, American Mathematical Society, 201 Charles St., Providence, RI 02904-2294 USA.

Editors

Titles in This Series

900 **Wolfgang Bertram,** Differential geometry, Lie groups and symmetric spaces over general base fields and rings, 2008

899 **Piotr Hajłasz, Tadeusz Iwaniec, Jan Malý, and Jani Onninen,** Weakly differentiable mappings between manifolds, 2008

898 **John Rognes,** Galois extensions of structured ring spectra/Stably dualizable groups, 2008

897 **Michael I. Ganzburg,** Limit theorems of polynomial approximation with exponential weights, 2008

896 **Michael Kapovich, Bernhard Leeb, and John J. Millson,** The generalized triangle inequalities in symmetric spaces and buildings with applications to algebra, 2008

895 **Steffen Roch,** Finite sections of band-dominated operators, 2008

894 **Martin Dindoš,** Hardy spaces and potential theory on C^1 domains in Riemannian manifolds, 2008

893 **Tadeusz Iwaniec and Gaven Martin,** The Beltrami Equation, 2008

892 **Jim Agler, John Harland, and Benjamin J. Raphael,** Classical function theory, operator dilation theory, and machine computation on multiply-connected domains, 2008

891 **John H. Hubbard and Peter Papadopol,** Newton's method applied to two quadratic equations in $\mathbb{C}^2$ viewed as a global dynamical system, 2008

890 **Steven Dale Cutkosky,** Toroidalization of dominant morphisms of 3-folds, 2007

889 **Michael Sever,** Distribution solutions of nonlinear systems of conservation laws, 2007

888 **Roger Chalkley,** Basic global relative invariants for nonlinear differential equations, 2007

887 **Charlotte Wahl,** Noncommutative Maslov index and eta-forms, 2007

886 **Robert M. Guralnick and John Shareshian,** Symmetric and alternating groups as monodromy groups of Riemann surfaces I: Generic covers and covers with many branch points, 2007

885 **Jae Choon Cha,** The structure of the rational concordance group of knots, 2007

884 **Dan Haran, Moshe Jarden, and Florian Pop,** Projective group structures as absolute Galois structures with block approximation, 2007

883 **Apostolos Beligiannis and Idun Reiten,** Homological and homotopical aspects of torsion theories, 2007

882 **Lars Inge Hedberg and Yuri Netrusov,** An axiomatic approach to function spaces, spec tral synthesis and Luzin approximation, 2007

881 **Tao Mei,** Operator valued Hardy spaces, 2007

880 **Bruce C. Berndt, Geumlan Choi, Youn-Seo Choi, Heekyoung Hahn, Boon Pin Yeap, Ae Ja Yee, Hamza Yesilyurt, and Jinhee Yi,** Ramanujan's forty identities for Rogers-Ramanujan functions, 2007

879 **O. García-Prada, P. B. Gothen, and V. Muñoz,** Betti numbers of the moduli space of rank 3 parabolic Higgs bundles, 2007

878 **Alessandra Celletti and Luigi Chierchia,** KAM stability and celestial mechanics, 2007

877 **María J. Carro, José A. Raposo, and Javier Soria,** Recent developments in the theory of Lorentz spaces and weighted inequalities, 2007

876 **Gabriel Debs and Jean Saint Raymond,** Borel liftings of Borel sets: Some decidable and undecidable statements, 2007

875 **C. Krattenthaler and T. Rivoal,** Hypergéométrie et fonction zêta de Riemann, 2007

874 **Sonia Natale,** Semisolvability of semisimple Hopf algebras of low dimension, 2007

873 **A. J. Duncan,** Exponential genus problems in one-relator products of groups, 2007

872 **Anthony V. Geramita, Tadahito Harima, Juan C. Migliore, and Yong Su Shin,** The Hilbert function of a level algebra, 2007

871 **Pascal Auscher,** On necessary and sufficient conditions for L^p-estimates of Riesz transforms associated to elliptic operators on $\mathbb{R}^n$ and related estimates, 2007

870 **Takuro Mochizuki,** Asymptotic behaviour of tame harmonic bundles and an application to pure twistor D-modules, Part 2, 2007

869 **Takuro Mochizuki,** Asymptotic behaviour of tame harmonic bundles and an application to pure twistor D-modules, Part 1, 2007

868 **Gelu Popescu,** Entropy and multivariable interpolation, 2006

867 **Vilmos Totik,** Metric properties of harmonic measures, 2006

866 **William Craig,** Semigroups underlying first-order logic, 2006

865 **Nathanial P. Brown,** Invariant means and finite representation theory of $C*$-algebras, 2006

864 **John M. Lee,** Fredholm operators and Einstein metrics on conformally compact manifolds, 2006

863 **M. Lübke and A. Teleman,** The Universal Kobayashi-Hitchin correspondence on Hermitian manifolds, 2006

862 **Alberto Canonaco,** The Beilinson complex and canonical rings of irregular surfaces, 2006

861 **Leon A. Takhtajan and Lee-Peng Teo,** Weil-Petersson metric on the universal Teichmüller space, 2006

860 **Thomas M. Fiore,** Pseudo limits, biadjoints and pseudo algebras: Categorical foundations of conformal field theory, 2006

859 **N. Arcozzi, R. Rochberg, and E. Sawyer,** Carleson measures and interpolating sequences for Besov spaces on complex balls, 2006

858 **Enrico Valdinoci, Berardino Sciunzi, and Vasile Ovidiu Savin,** Flat level set regularity of p-Laplace phase transitions, 2006

857 **Donatella Danielli, Nocola Garofalo, and Duy-Minh Nhieu,** Non-doubling Ahlfors measures, perimeter measures, and the characterization of the trace spaces of Sobolev functions in Carnot-Carathéodory spaces, 2006

856 **Vladimir Bolotnikov and Harry Dym,** On boundary interpolation for matrix valued Schur functions, 2006

855 **Yevgenia Kashina, Yorck Sommerhäuser, and Yongchang Zhu,** On higher Frobenius-Schur indicators, 2006

854 **Noam Greenberg,** The role of true finiteness in the admissible recursively enumerable degrees, 2006

853 **Joachim Krieger,** Stability of spherically symmetric wave maps, 2006

852 **Viorel Barbu, Irena Lasiecka, and Roberto Triggiani,** Tangential boundary stabilization of Navier-Stokes equations, 2006

851 **Jie Wu,** On maps from loop suspensions to loop spaces and the shuffle relations on the Cohen groups, 2006

850 **Siegfried Echterhoff, S. Kaliszewski, John Quigg, and Iain Raeburn,** A categorical approach to imprimitivity theorems for C^*-dynamical systems, 2006

849 **Katsuhiko Kuribayashi, Mamoru Mimura, and Tetsu Nishimoto,** Twisted tensor products related to the cohomology of the classifying spaces of loop groups, 2006

848 **Bob Oliver,** Equivalences of classifying spaces completed at the prime two, 2006

847 **Eric T. Sawyer and Richard L. Wheeden,** Hölder continuity of weak solutions to subelliptic equations with rough coefficients, 2006

846 **Victor Beresnevich, Detta Dickinson, and Sanju Velani,** Measure theoretic laws for lim–sup sets, 2006

For a complete list of titles in this series, visit the AMS Bookstore at **www.ams.org/bookstore/**.